Techniques of
Thin-Layer Chromatography
in Amino Acid and Peptide
Chemistry

Techniques of Thin-Layer Chromatography
in Amino Acid and Peptide Chemistry

Second English Edition (Revised)

By György Pataki

Laboratory for Chromatography
Analytical Department
Robapharm Ltd.
Basel

Translated from the original German
by STS of
Ann Arbor, Michigan

With 127 figures, 52 tables,
497 literature references, and
a bibliography of new applications
containing 489 additional references.

 ANN ARBOR – HUMPHREY SCIENCE PUBLISHERS, INC.
Ann Arbor and London

Foreword

Progress in the natural sciences is impossible without the expansion and refinement of analytical methods. After the early monopoly held by the balance and test tube, expansion was achieved by the use of new physical instruments—up to our modern mass spectrometer; and after the introduction of microanalysis by Pregl, refinements were made by advances in the area of nanograms, which was opened by modern chromatographic methods.

In this continuously more rapid growth, the greatest benefits have probably been reaped by the chemistry and biochemistry of amino acids, peptides, and proteins. Paper chromatography and paper electrophoresis were developed on this ground. In a short time, however, this important field availed itself of the services of thin-layer chromatography and did so to such an extent that the publication of special instructions is justified and necessary only five years later.

Anyone who today wishes to identify known or unknown amino acids in living cells, to analyze the amino acid composition of a peptide, to compare proteins of different origin in the "fingerprint" on the basis of enzymatic cleavage peptides, or to verify the results of his peptide syntheses, must be familiar with thin-layer techniques in chromatography and electrophoresis.

The present laboratory handbook is intended as a guide and reference to this end.

Frankfurt/Main   THEODOR WIELAND

Preface to the
Second English Edition
(Second Revision)

Less than one year after the first English edition has been published, there is a need for reprinting the book, since the first printing is depleted. Since the time to supply the demand was short, it has been decided not to revise the text of the first edition. On the other hand, several hundred additional references as well as a Bibliography of New Applications of TLC have been added in order to make the book more up-to-date. In this way, the second (revised) English edition contains all the references which came to the author's attention until September, 1968. It is hoped that this revised and supplemented second edition will be of value for workers in amino acid, peptide and protein fields.

Special acknowledgment is due to Mrs. C. Calcagno for her valuable help in preparation and correction.

GYÖRGY PATAKI

Riehen (Basel), April 1969

Preface to the
First English Edition
(Revised)

Since the first German edition has been published, a great number of papers have appeared on the subject of thin-layer chromatography in general as well as on the application in amino acid and peptide chemistry. In this edition a selection of the most important papers have been included, in order to increase the value of the book. The author hopes that this supplemented edition will be useful.

Special acknowledgment is due to Dr. U. P. Geiger for his valuable help in preparation and correction.

GYÖRGY PATAKI

Riehen (Basel), April 1967

Preface

Thin-layer chromatography (TLC) was developed by E. Stahl into an indispensable laboratory method. Numerous publications in nearly all fields have shown that, in many cases, this technique is superior to paper chromatography. Although it was anticipated that TLC would not be very suitable for the separation of hydrophilic substances (e.g. amino acids, peptides, and sugars), intensive development began in recent years toward analysis of these compounds and its end is not yet in sight. Publications in TLC in the field of amino acid and peptide chemistry for 1958 to 1964 have increased exponentially. This extraordinary progress would indicate that even a handbook prepared by specialists will hardly be able to keep step with future development.

In the present monograph, an attempt is made to offer a general survey concerning the application of TLC for the analysis of amino acids, peptides, and related compounds. Approximately one third of the book is devoted to general problems of TLC, with an intentional neglect of completeness in details. The first part deals with the technique of TLC to such an extent that the reader can prepare and evaluate the chromatograms without the aid of the original literature. The literature data available on the subjects of this special part are very numerous, and in part are published in periodicals that are not readily accessible. Parts II, III, and IV, with few exceptions, discuss only those studies which have appeared prior to November, 1964. In addition, many unpublished findings are presented which were generously placed at the disposal of the author or were obtained in his own laboratory.

Many colleagues have contributed to the success of this book by making their work available prior to publication and by reporting their unpublished findings. It is a pleasant duty to take this opportunity to express my sincere appreciation to Mrs. El. Bujard, Mrs. F. Marcucci, Mrs. J. Opienska-Blauth, Prof. of Chemistry, and Messrs. Dr. J. Barrolier, Prof. M. Behrens, Prof. C. H. Brieskorn, Prof. J. M. Buchanan, Dr. W. Bürgi, Dr. H. C. Curtius, Dr. J. Dittmann, Dr. H. W. Goeede, Prof. L. Hörhammer, Dr. B. Lerch, E. Mussini, Dr. A. Niederwieser, Dr. R. S. Ratney, Dr. T. Rokkones, Dr. E. Schröder, Dr. N. Seiler, Dr. H.

Stegemann, Dr. F. Tancredi, Prof. Theodor Wieland, Dr. B. Witkop and G. Wohlleben.

I am indebted to Prof. Th. Koller and Dr. M. Keller for their generous support of biochemical research which has made the present monograph possible.

Special acknowledgment is also due to Mrs. Anneliese Geiger-Cloos, Mrs. E. Pataki-Lehnen, M. Pharm., and Mr. U. P. Geiger, B.Sc., for their aid in searching and evaluating the literature and in preparing and correcting the manuscript. For many of the illustrations, I am grateful to Miss H. Hertig and Mssrs. Dr. H. R. Bolliger, W. Bürki and F. Waldmeyer. It is hoped that this book will stimulate chemists, biochemists, biologists, pharmacists, and the medical profession and that it will be an aid in the practical performance of their work.

GYÖRGY PATAKI

Contents

INTRODUCTION   xvii

Part 1. **Technique of Thin-Layer Chromatography**

1. Preparation of the Layer   3

  A. Applicators   3
     1. Applicators of the Stahl type   3
     2. Kirchner-Type Applicators   5
     3. Simple Devices   6
     4. The Microplate Method   7

  B. Manual Layer Preparation   8

2. Development of Chromatograms   10

  A. Application of Test Substances   10
  B. Ascending Development   12
     1. Rectangular Separation Chamber   12
     2. The Covered Plate   14
  C. Horizontal Development   15
  D. Multiple Development   16
  E. Continuous Flow Chromatography and
     Wedge-Strip Technique   19
     1. Continuous Flow Development in the BN Chamber   20
       a. Variant A   20
       b. Variant B   20
     2. Other Continuous Flow Methods   20
     3. Wedge-Strip Technique   21
  F. Multidimensional Chromatography   21
     1. Two-Dimensional Chromatography and
       Separation-Reaction-Separation Technique   21
     2. Multidimensional Chromatography   22
  G. Polyzonal Thin-Layer Chromatography   23
  H. Gradient Development   25
     1. Gradient Elution   26
     2. Chromatography on Gradient Layers   28
  I. Preparative Thin-Layer Chromatography   28
     1. Relations to Column Chromatography   28
     2. Chromatography on Thick Layers   30
  J. Thin-Layer Electrophoresis   33

K. Radiochromatography   34
   1. Autoradiography   35
   2. Evaluation with Counter Tube   35
   3. Evaluation with Scintillation Counter   36
L. Instructions for the Preparation of Thin-Layer
   Chromatograms   38

3. Evaluation of the Chromatograms   39

A. R_f-Values in Thin-Layer Chromatography   40
   1. Reproducibility of R_f-Values, Factors of Influence   40
      a. Activity of the Layer   40
      b. Layer Thickness, Plate Format, and Layer Preparation   41
      c. Chamber Saturation   41
      d. Chromatographing Technique   43
      e. Development Distance and Distance of
         Starting Point from the Surface of the Solvent   43
      f. Quality of the Solvent   43
      g. Quantity of Substance and Solvent   43
      h. Temperature   45
      i. Contaminants   45
   2. Consequences in Practice   46
   3. Comparison with Paper Chromatography   48
B. R_f-Value and Chemical Structure   49
   1. The Martin Relation   49
      a. Results of Thin-Layer Chromatography   50
      b. Deviations from the Martin Relation   51
      c. The Martin Relation in the Case of
         Chromatographic Solvent Demixing   53
   2. Applications of the Martin Relation   55
C. Quantitative Thin-Layer Chromatography   56
   1. In Situ Evaluation   57
   2. Evaluation after Elution   60

Part 2. **Thin-Layer Chromatography of Amino Acids,**
        Peptides, and Related Compounds

4. Thin-Layer Chromatography of Amino Acids   65

A. Chromatography on Silica Gel Layers   65
   1. Preparation of the Layer   65
   2. Solvents and R_f-Values   66
   3. Separation of Amino Acids   71
B. Chromatography on Other Inorganic Layers   74
   1. Aluminum Oxide and Celite Layers   74
   2. Mixed Silica Gel/Kieselguhr Layers   76
C. Chromatography on Cellulose Layers   76
   1. Preparation of the Layer   76
   2. Solvents and R_f-Values   79
   3. Two-Dimensional Separation of Amino Acids   79
   4. Multidimensional Separation of Amino Acids   81

D. Electrochromatography   85
E. Detection of Amino Acids   88
   1. The Ninhydrin Reaction   90
      a. Silica Gel Layers   90
      b. Cellulose Layers   90
      c. Polychromatic Detection   92
   2. Non-Destructive Detection   92
   3. Other Color Reactions   93
      a. Chlorine/Tolidine Reaction   93
      b. Isatin Reaction   93
      c. Paulys Reagent   93
      d. Folin Reagent   94
      e. Sodium Nitroprusside/
         Potassium Ferricyanide Reagent   94
      f. Sakaguchi Reaction   94
F. Thin-Layer Chromatography of Iodoamino Acids   94
   1. Solvents and Separation Effects   94
   2. Detection of Iodoamino Acids   97

5. Peptides and Intermediates of Peptide Synthesis   98

A. Thin-Layer Chromatography of Peptides   98
B. Control of Peptide Syntheses   102
C. Detection of Peptides and Peptide Derivatives   107
   1. Chlorine/Tolidine Reaction   107
      a. Procedure According to Brenner et al.   107
      b. Procedure According to von Arx and Neher   107
      c. Procedure According to Barrolier   108
   2. Other Color Reactions   108
      a. Iodine/Starch Reaction   108
      b. Detection with Morin   108
      c. Detection with Chromic Sulfuric Acid   108
      d. Detection of Cysteine and Cystine Derivatives   108
      e. Detection of Hydrazides,
         Aminodiacylhydrazines, and Diacylhydrazines   109

**Part 3.   Application of Thin-Layer Chromatography in
the Sequential Analysis of Proteins and Peptides**

6. Thin-Layer Chromatographic Analysis of Protein and
Peptide Hydrolysates   113

A. Amino Acids in Total Hydrolysates   114
   1. Acid Hydrolysis   114
   2. Alkaline Hydrolysis   115
   3. Hydrolysis with Ion Exchangers   116
   4. Applications   116
B. Thin-Layer Fingerprint Technique   121

7. N-(2,4-Dinitrophenyl)-Amino Acids   126

A. Dinitrophenylation   126

1. Amino Acids   127
   a. Preparation of Dinitrophenyl-Amino Acids   127
   b. Dinitrophenylation of an Amino Acid Mixture   128
2. Peptides   129
   a. Conversion with Dinitrofluorobenzene   129
   b. Total Hydrolysis of a Dinitrophenyl Peptide   129
3. Polypeptides and Proteins   130
   a. Conversion with Dinitrofluorobenzene   130
   b. Total Hydrolysis of a Dinitrophenyl Protein   130
   c. Partial Hydrolysis of a Dinitrophenyl Protein   131

B. Thin-Layer Chromatography of
   Water-Soluble Dinitrophenyl-Amino Acids   131
C. Thin-Layer Chromatography of
   Ether-Soluble Dinitrophenyl-Amino Acids   132
   1. Solvents and R_f-Values   132
   2. One- and Multidimensional Separation   137
D. Detection Methods   141
   1. Photocopy   142
   2. Photography   142

8. N-(2,4-Dinitro-5-Aminophenyl)- and 1-Dimethylamino-
   naphthalene-5-Sulfonyl Amino Acids   144

A. Preparation   144
   1. Dinitroaminophenyl-Amino Acids   144
   2. 1-Dimethylaminonaphthalene-5-Sulfonyl Amino-Acids   145
B. Determination of the
   N-Terminal Amino Acids of a Peptide   145
C. Chromatography   147
   1. Solvents and R_f-Values   147
   2. One- and Two-Dimensional Separation   147
D. Methods of Detection   147

9. 3-Phenyl-2-Thiohydantoins   149

A. Preparation of PTH-Amino Acids   149
   1. Microsynthesis According to Sjöquist   150
   2. Synthesis According to Edman   150
   3. Synthesis According to Cherbuliez   150
   4. Quantitative Conversion of an
      Amino Acid Mixture into PTH-Derivatives   151
B. Degradation of Peptides and Proteins   151
   1. Stepwise Degradation of Peptides   153
      a. Peptide Degradation According to Sjöquist   153
      b. Peptide Degradation According to Sjöholm   154
      c. Peptide Degradation According to Schroeder   154
      d. Peptide Degradation According to Wieland   155
      e. Radioisotope Method of Cherbuliez   155
   2. Degradation of Proteins   156
      a. Protein Degradation
         According to Ericksson and Sjöquist   156
      b. Radioisotope Method According to Laver   156

C. Chromatography of PTH-Amino Acids   157
  1. Solvents and Separation Effects   157
  2. Determination of the N-Terminal Amino
     Acid and Degradation of Peptides   160
  3. Detection Methods   163
     a. Chlorine/Tolidine Reaction   163
     b. Butylhypochloride/KI/Starch Reaction   163
     c. Iodine Azide Reaction   ł63
     d. UV-Light   163
     e. Radioisotope Method   163
     f. Specific Detection of PTH-Glycine   163
10. Trinitrophenyl Amino Alcohols   164

**Part 4. Amino Acids and Related Compounds in
         Biological Material**

11. Chromatography and Electrochromatography of Free
    Amino Acids   169
  A. Preparations for Chromatography   169
     1. Removal of Proteins, Polysaccharides, and Lipoids   170
     2. Decomposition of Urea   171
     3. Desalting   171
     4. Determination of α-Amino Nitrogen   172
  B. Separation of Amino Acids   173
     1. Amino Acids in Urine and Blood   173
     2. Additional Applications   179
12. Chromatography of Dinitrophenyl Amino Acids   182
  A. Preparation for Chromatography   182
     1. Dinitrophenylation   182
        a. Amino Acids in Urine   182
        b. Amino Acids in Blood   183
        c. Amino Acids in Sperm   183
     2. Extraction of DNP-Amino Acids   183
        a. Ether-Soluble DNP-Amino Acids   183
        b. Water-Soluble DNP-Amino Acids   183
  B. Separation of DNP-Amino Acids   184
     1. Amino Acids in Urine, Blood, and Sperm   184
     2. Other Applications   190

References   193
Bibliography of New Applications   211
Scheme I   245
Schemes II & III   247
Commercial Suppliers   248
Index   249

Introduction

The discovery of paper chromatography (Consden, Gordon, and Martin) undoubtedly represents a great advance in protein chemistry. A high point in this development was reached with the determination of the constitution of insulin (Sanger).

Thin-layer chromatography (TLC) was described in 1938 by the Russian authors Ismailov and Shraiber [180], but, similar to the case of Tswett's adsorption chromatography, it was soon forgotten. Later, the method reappeared by the name of "surface chromatography" [250], "chromatostrips" [199], and "chromatoplates" [337]. The adsorbent was usually applied on the glass support in aqueous suspension; Meinhard and Hall [250] were the first to introduce starch as a binding agent.

A fundamentally different route was taken by Crove [75] and others [443, 262]. They dusted the dry adsorbent on glass plates ("spread-layer chromatography") and carried out the separation on this loose layer.

Chromatography on thin adhesive layers has many advantages over the dusting method and was subsequently used by many investigators, especially for the analysis of terpenes (summary in [238]). Surprisingly, however, neither the organic chemist nor the biochemist gave early recognition to TLC. Only with the development of a spreader and the introduction of a suitable adsorbent (Silica Gel G) was it possible for Stahl [409, 405, 407, 408] to make this elegant method of microanalysis generally applicable.

Originally it was assumed that TLC was exclusively suited for the analysis of lipophilic substances. In 1958-62 it was found, however, that amino acids, among other substances, could be chromatographed in thin layers. The first systematic investigation on the separation of amino acids was carried out by Brenner and Niederwieser [42]. They used Silica Gel G as the adsorbent. Such an attempt seemed indicated, since the older literature contains a larger number of data concerning column chromatography of amino acids on inorganic adsorbents.

Compared to paper chromatography, thin-layer chromatography has the following advantages: (a) the limits of detection are lowered, (b) the running time is considerably shorter, and (c) the separation is sharper (see Fig. 1). All of these advantages reside in the small amount of spreading of the spots of test substance. Thus, for example, 14 amino acids can easily be separated on a Silica Gel G layer; under identical conditions, 10 amino acids are separated only incompletely on Whatman No. 1 paper.

It can be claimed today that TLC is the method of choice for the separation of amino acids and related compounds.

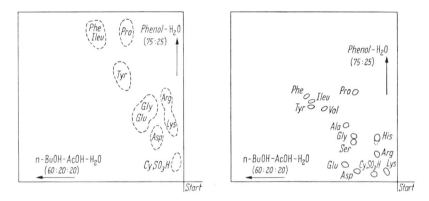

Fig. 1. Comparison between paper and thin-layer chromatography. (a) 10 Amino acids on Whatman No. 1 paper (running time 2 and 2½ hrs., respectively). (b) 14 Amino acids on Silica Gel G layer (running time 1½ and 2 hrs., respectively). Amount of substance applied: 1 μg amino acid in 1 μl water in each case. According to Brenner and Niederwieser [42].

PART I

Technique of
Thin-Layer
Chromatography

Chapter 1

Preparation of the Layer

The uniformity of an adsorption layer is an essential prerequisite for the separation as well as for the reproducibility of the results. Both the separation of the mixture of substances and the reproducibility depend upon the preparation of the layer. The layer can be prepared with or without an applicator.*

A. APPLICATORS

The applicators which are customarily used today can be classified into two groups: applicators of the *Stahl* type [405, 410] and applicators of the *Kirchner* type [252]. According to Stahl, the layer is prepared with a movable application device; Kirchner uses a fixed device.

1. Applicators of the Stahl Type

The Stahl applicator [405, 410] consists of two parts: working template and applicator. Glass plates of *equal thickness* are placed on the template in a row. The applicator holds about 150 ml liquid. The layer thickness can be varied between 0-2 mm. Fig. 2 shows the device and the procedure.†

To fill the applicator, the lever must point in the direction of the red arrow engraved on its top. After transferring the suspension, the lever is turned 180° to the left and spreading begins only after the suspension has visibly appeared.

A simplified design‡ is shown in Fig. 3.

---

* Recently, thin-layer plates which are ready for use have become available on the market. (Custom Service Chem., Inc., Wilmington, Del.; Distillation Products, Inc., Division of Eastman Kodak Co., Rochester, Minn.; Gelman Instrument Co., Ann Arbor, Mich.).

† C. Desaga GmbH, Heidelberg, Germany.

‡ Research Specialties Co., Richmond, Calif.

With the use of the Unoplan applicator,* it is also possible to coat glass plates of *different thickness* (Fig. 4). By compressing the rubber ball several times (Fig. 4), an air cushion is pumped under the roll. This forces the glass plates against the underside of the side guide rail. In this manner, the surfaces of all plates are brought to one level.

A semiautomatic device has been described by Marcucci and Mussini [243] (compare also [405, 409, 410]).

Fig. 2. Coating of plates with the Stahl applicator (Desaga GmbH, Heidelberg, Germany).

---

* Shandon Scientific Co., London, England.

Fig. 3. Simplified applicator without tipping mechanism (Research Spec. Co., Richmond, Cal.).

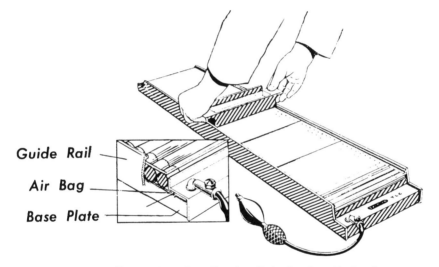

Guide Rail

Air Bag

Base Plate

Fig. 4. Applicator (Unoplan, Shandon Ltd., London, England).

2. Kirchner-Type Applicators

The spreading principle described by Kirchner [252] recently drew the attention of Wollisch [448]. Fig. 5 shows an applicator of the Camag Company.*

Fig. 5. Applicator (Camag, Muttenz/ Bl., Switzerland). *Photo—courtesy Gelman Instrument Company.*

---

* Camag, Muttenz/Bl., Switzerland.

A guide rail is screwed on the baseboard. The side walls of the chamber for the adsorbent suspension are movable. The layer thickness can be adjusted between 0-5 mm. For coating purposes, the suspension (120-160 ml) is filled into the chamber and the first plate, which has already been introduced, is uniformly advanced by the succeeding plate. *If plates of different thickness are coated, they must be pushed through in the sequence of decreasing thickness.*

Camag has developed a semiautomatic device also (Fig. 6). The plate transport is automatic.*

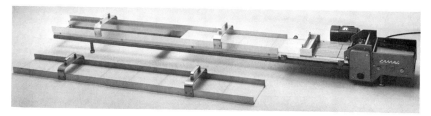

Fig. 6. Automatic applicator (Camag, Muttenz/Bl., Switzerland). *Photo—courtesy Gelman Instrument Company.*

3. Simple Devices

The applicators described above can be replaced by simpler, do-it-yourself devices. Barbier et al. [9] prepare layers of 0.3 mm thickness with a simple Plexiglas applicator (Fig. 7).

The applicator of Lüdy-Tenger [233] also is simple to build; the author recommends the use of photographic plates of 9 × 12 cm for coating.

Each of these devices works according to the principle of Stahl.

The applicator suggested by Kratzl [211] consists of a glass template and a beveled glass ruler (Fig. 9).

For the preparation of the template, microscope cover glasses (C) are

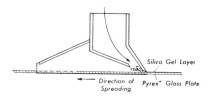

Fig. 7. Simple Plexiglas applicator according to Barbier et al. [9].

Fig. 8. Simple applicator according to Lüdy-Tenger [233].

---

* Additional Kirchner type applicators can be obtained from Sas & Sas Ltd., London, England; Townson & Mercer, Beddington Lane, Coryden-Surrey, England.

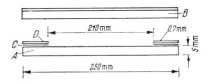

Fig. 9. Work template and applicator
suggested by Kratzl [211] (compare
text).

cemented at a distance of about 5 mm on both longitudinal sides of
a glass plate (A, 25 × 30 cm from glass of 5 mm thickness), and a glass
strip (D, 2 × 30 cm and 0.7 mm thick) is applied on each of the cover
glasses. For the preparation of a layer of 0.3 mm thickness, the plate is
placed on the somewhat tilted template. The suspension is poured on
the upper edge and drawn down with the ruler (B). A similar method
is also used by Duncan [101], as well as by Lees and De Muria [220].

The preparation of non-adhesive adsorbent layers is described by
Cerny et al. [61], Mottier [261], and Davidek [83]. Fig. 10 shows the
method of Davidek [83].*

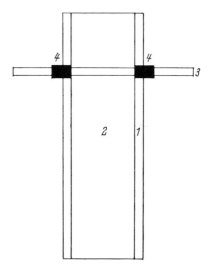

Fig. 10. Preparation of loose non-
adhering layers according to Davidek
[83].
   1—plate glass
   2—polyamide powder, spread with
      roll
   3—roll for smoothing of layer
   4—rubber tubing

Some authors prepare the layer by spraying of the adsorbent suspension [21a, 323, 428].

4. The Microplate Method

It is known from paper chromatography that a simple relation
$(Z_f)^2 = a + b \cdot t$ exists between the migration rate of the front (Z_f) and
the running time (t) [350, 263, 209, 137, 449]. This relation was also

---

   * Recently, a complete kit for chromatography by the dusting method has
become available (Aimer Prod. Ltd., London, England).

confirmed in TLC [272, 8]. As indicated by Fig. 11, the running time can be reduced considerably by decreasing the development distance.

For the preparation of small plates, the Stahl applicator (p. 4) is used by Prey et al. [8]. Twenty-four small ($75 \times 75 \times 4$ mm) glass plates are coated with the adsorbent in one step. The Desaga Company has developed a small basic kit for TLC (Fig. 12) which includes an applicator for the preparation of small plates.

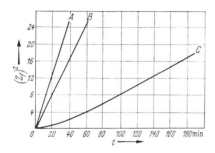

Fig. 11. Relation between migration rate of the front and running time (horizontal technique); p. 15.
A—ether-glacial acetic acid (95 : 5)
B—chloroform
C—phenol-water (75 : 25 g/g)
From Niederwieser [272].

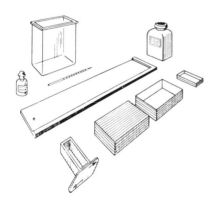

Fig. 12. Micro-kit for thin-layer chromatography with applicator for small plates and for ribbed glass (Desaga GmbH, Heidelberg, Germany).

This device permits the coating of extremely inexpensive commercial plates [130]. Recently, the Camag Company has also introduced a kit for the preparation of small plates.

B. MANUAL LAYER PREPARATION

Although a rational preparation of thin layers of uniform thickness is possible only with the use of a suitable applicator, it has been proposed occasionally to prepare the plates manually. This method of layer preparation may furnish more or less uniform layers after sufficient practice, but the reproducibility leaves something to be desired (p. 41). With the use of manually prepared layers it must be recommended, in any case, that the chromatogram be monitored by the application of dye mixtures at different spots of the plate. Hörhammer et al. [169] have suggested that the adsorbent be suspended in ethylacetate instead of water; Bhandari et al. [29] use a mixture of ethyl alcohol and water (the latter procedure is claimed to have an unfavorable influence on the separation; p. 30).

The suspension is poured on the plate in such a manner that its surface is half covered (Fig. 13).

Fig. 13. Manual layer preparation: pouring of the suspension. From Hörhammer et al. [169].

Fig. 14. Manual layer preparation: spreading of the adsorbent. From Hörhammer et al. [169].

The plate is now placed on the palm of one hand, and the suspension is allowed to flow slowly along the edges and into the corners by repeated careful tilting until the entire surface has been coated (Fig. 14).

In this procedure, the plate may not be held by the edges because the suspension may run over the edge of the plates. Subsequently, the layer is allowed to dry in air in *horizontal* position.

Ribbed glass [130] can also be coated manually. The layer thickness is defined by the depth of individual grooves. The highly original method of Feltkamp [112]* is particularly suitable for preliminary experiments. Glass rods with a length of 200 mm and diameter of 7.5 mm serve as the carriers of the adsorbent. A number of cork stoppers with a bore fit into test tubes of 18 × 18 mm, and the glass rods are held in them but remain movable. The glass rods should project 20-30 mm over the top of the cork. The suspension is filled into a vertical tube with a base that is designed in a way that a glass rod just fits in. The rods are immersed so that they remain dry about 10 mm below the cork. The rods are suspended vertically for drying. Cseh [76], who used water as the suspension medium, observed that the layer is easily detached from the rods.

---

* F. Buhler, Tübingen, Germany.

Chapter 2

Development of Chromatograms

A. APPLICATION OF TEST SUBSTANCES

Free amino acids and peptides are hydrophilic compounds and, consequently, are sparingly soluble in non-aqueous solvents. For the determination of R_f values (for definition see p. 39) or for the preparation of two-dimensional spot cards (p. 21), a collection of pure amino acids* is used and solutions are prepared with them which contain 1 mg/ml of the substances to be tested. A solvent consisting of 10% isopropyl alcohol is suitable, for example.†

The free amino acids are present as hydrochlorides in hydrochloric acid protein hydrolysates. In such cases, it is also recommended to apply reference substances in hydrochloric acid solution (1 mg amino acid in 1 ml 0.1 N HCl). Before chromatography, the solvent must be completely removed (ventilation for 10-20 min.). In paper chromatography, it is customary to neutralize the HCl excess by ammonia vapors. This procedure is not recommended with Silica Gel G layers, since ammonia adheres relatively firmly to silica gel and may interfere with development.

Amino acids in organ extracts or body fluids must be freed from contaminants (p. 169) or must be chromatographed in the form of suitable derivatives (p. 182).

A substitution of the amino acids at the amino group, at the carboxyls, or at both converts these compounds into acids, bases, or neutral substances. The zwitterionic character is thus lost. Depending upon the type of substituant, the derivatives are always still more or less polar. Most amino acid derivates (dinitrophenyl-, dinitroaminophenyl-amino acids,

---

* Collections are supplied by the following companies: Fluka, Buchs/SG., Switzerland; Merck AG, Darmstadt, Germany; Cyclo Chemical Company, Los Angeles, Calif.; California Corp Biochem. Research, Los Angeles, Calif.; Mann Research Co., New York, N.Y.; Nutritional Biochem. Co., Cleveland, Ohio.

† A collection of 10^{-2} M amino acid solutions is available from Shandon Ltd.

phenylthiohydantoins) result in a loss of zwitterionic character and are partially insoluble in water. Consequently, they are applied in suitable organic solvents. (For the preparation and suppliers of these compounds as well as the preparation of test solutions, see pp. 126 ff, 147, 160-161.)

For preliminary experiments and qualitative studies, it is possible to use home-made capillaries. With some practice, the applied quantity can be estimated from the size of the wetted area. Lüdy-Tenger [233] proposed the use of platinum loops. The use of graduated micropipettes (Desaga) or an applicator set (Camag) permits somewhat more accurate work.

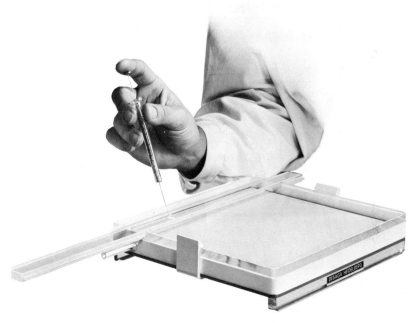

Fig. 15. Preparation tray for the application of sensitive substances with gas blanket; Hamilton metering syringe (Desaga GmbH, Heidelberg, Germany).

For quantitative determinations, the "Drummond Microcaps" are excellent.* They are to be recommended especially in sizes of 1-10 μl. Microsyringes (Hamilton or Agla syringe—Desaga) are characterized by high precision; however, they are relatively expensive and difficult to handle in series studies. Fig. 15 shows a preparation box by means of which sensitive substances can be applied under an N_2 blanket.

The mixtures to be separated and the reference samples are applied at a distance of 1.5 cm from the plate edges. If the volume of sample does

---

* P. Sütterlin, Dietikon/ZH., Switzerland; Shandon Ltd.

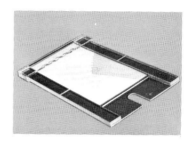

a. Desaga GmbH, Heidelberg,   b. Camag AG, Muttenz/Bl.,
    Germany.   Switzerland.

Fig. 16. Spotting template.

not exceed 1 μl, a spacing of about 0.8 cm is permitted. In that case, there is room for about 20 samples on a 20 × 20 cm plate. The samples can be applied in a line by means of the very practical application template (Fig. 16).

For the band application of substances in preparative chromatography, see p. 31.

B. ASCENDING DEVELOPMENT

1. Rectangular Separation Chamber

As a rule, glass plates of 20 × 20 cm serve for chromatography. Development takes place most suitably with "chamber saturation" in the rectangular separating chambers produced especially for TLC (with regard to chamber saturation, see p. 41).

Fig. 17 shows the Desaga tank; for the chromatography of sensitive substances, it can be provided with a special lid through which nitrogen, for example, can be introduced into the chamber.

The rectangular tanks of the Shandon Company can also be arranged for plates of 5 × 20 cm size. This has the advantage that a number of one-dimensional amino acid chromatograms can be developed under identical conditions and can be sprayed with different dye reagents (Fig. 18).

If several 20 × 20 cm plates are to be placed into the same tank (this is recommendable particularly for the amino acid separation according to von Arx and Neher, p. 81), a bent blass rod can be used to hold the plates [42] (Fig. 19).

Another possibility is offered by the "chromatostack" technique, according to Nybom [280]. In this case, several coated plates are stacked (coated side against uncoated side), are covered with a clean plate, and

Fig. 17. Development tank for 20 × 20 cm plates. In foreground: Additional lid with gas feed line (Desaga GmbH, Heidelberg, Germany).

held together with rubber bands. To separate the individual plates from each other, plastic disks of 5 mm diameter are used and are placed on each plate corner. (This procedure actually is similar to the S-system; see below). For the simultaneous development of several chromatograms,

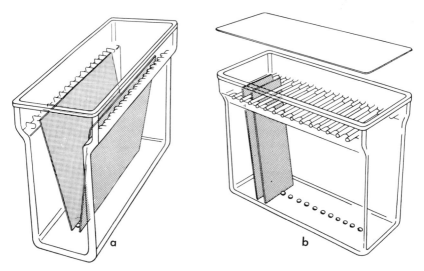

Fig. 18. Development tank. (a) Outfitted for 20 × 20 cm plates; (b) outfitted for 5 × 20 cm plates (Shandon Ltd., London, England).

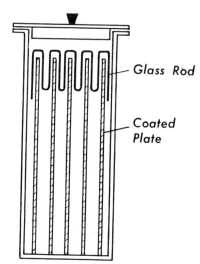

Glass Rod

Coated
Plate

Fig. 19. Development tank with simple glass holder for insertion of several plates. From Brenner and Niederwieser [42].

the Shandon Company has developed a special separating chamber (Fig. 20).

2. The Covered Plate

The chamber volume can be decreased if the carrier plate is used as the back wall of the chamber. A second glass plate, which is placed on the carrier plate, represents the other wall of the chamber. The principle of the "covered plate" was described by Brenner and Niederwieser [43] and later also by Stahl [410]. Fig. 21 shows the so-called S-chamber system.

The "20 cm S-chamber tank" serves to hold plates of 10 and 20 cm

Fig. 20. Separation tank for the simultaneous development of several plates (Shandon Ltd., London, England).

width, and the "40 cm S-chamber tank" is for 20 and 40 cm plates. A proper backing plate is available for each plate format. Simplifications of the S-system were proposed by Honegger [165], Davies [84], Wassicky [435] and Jänchen [183]. An inexpensive model is furnished by the Camag Company.

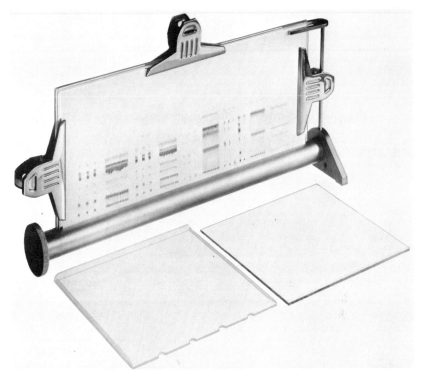

Fig. 21. S-chamber equipped for 40 × 20 cm plates. Foreground: cover plate (20 × 20 cm) and support plate (Desega GmbH, Heidelberg, Germany).

In some cases, the S-chamber system must be provided with additional "saturation" as described on p. 42. According to Honegger [165], the backing plate is provided with filter paper on the side turned toward the layer, and the paper is wetted with the solvent prior to the chromatography. Jänchen [183] uses a second coated plate as a backing.

C. HORIZONTAL DEVELOPMENT

Compared to ascending development, horizontal chromatography offers a few advantages particularly with the use of loose, non-adhering layers. A distinction is made between circular technique and horizontal chromatography.

Ismailov and Shraiber [180], as well as Meinhard and Hall [250], made use of a type of circular development and added the solvent in drops to the layer. The "microcircular technique" as described by Stahl [405, 470] is an excellent method for the determination of a suitable solvent. In the development of "spread-layer" chromatograms, the circular technique was also used [214, 215, 443]. The experimental arrangement of Stahl [407] served well for a horizontal circular chromatography.

The principle of the *covered plate* has been described on p. 14. A simple arrangement for horizontal chromatography was described by Brenner and Niederwieser [43] and by Hesse et al. [158].

Chromatography in the Desaga Brenner and Niederwieser (BN) chamber [272]: The solvent is transported to the thin-layer by the capillary action of the paper tongue. The cover plate prevents evaporation in the region of the separating path of 18 cm length. The cooling block prevents condensation of solvent vapors on the underside of the cover plate. Fig. 22 shows the BN chamber before installation of the plate.

The layer on two opposite edges of a coated glass plate (20×20 cm) is wiped off so that clean bands of about 6 mm width are formed. The samples are applied on the front edge at a distance of 15 mm from it, and the plate is placed on the cooling block while the front edge contacts the solvent tank (Fig. 23).

The tank is filled with about 20 ml solvent. A prescored paper tongue is bent about $90°$ along the fold line and is hung over the edge of the tank so that the folded end lies horizontally on the layer, and its open end dips vertically into the solvent (Fig. 23). It must be kept in mind that the solvent tank projects somewhat over the layer; if necessary, the springs must be slightly opened. Immediately afterward, the system is closed with the cover plate (Fig. 24). The zone of 20 mm width between the rims should just cover the tank (Fig. 24). The complete arrangement is shown in Figs. 23 and 24.

Finally, the cover plate is pressed down and fixed with the clamps. The bores are covered with a narrow glass plate. For additional applications of the BN chamber see pp. 19-20.

A great advantage of horizontal thin-layer chromatography in the BN chamber consists of the fact that the results are reproducible even in preparative separations (p. 34).

D. MULTIPLE DEVELOPMENT

In multiple chromatography, development takes place repeatedly and consecutively with the same solvent in the same direction, each time after intermediate drying. Although this procedure is often time-consuming, it has certain advantages to continuous-flow chromatography (p. 19);

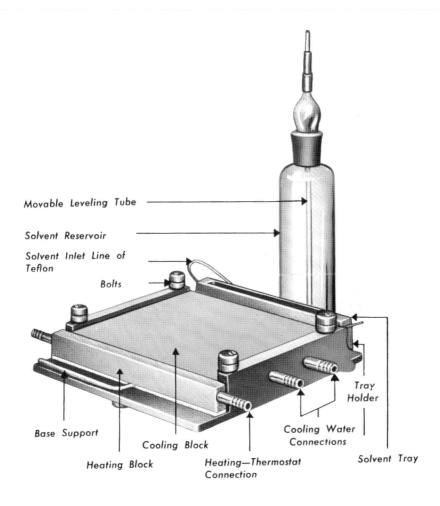

Fig. 22. BN chamber. General view before installation (Desaga GmbH, Heidelberg, Germany).

the spots usually spread less extensively. The R_f-value after the n-th run (^nR_f) is calculated as follows [184]:

$$^nR_f = 1 - (1 - R_f)^n$$

Diagrams for the determination of ^nR_f-values are found in the studies of Lenk [222] and Halpaap [150]. Information on the theory of one-dimensional multiple chromatography can be found in the work of Thoma [423].

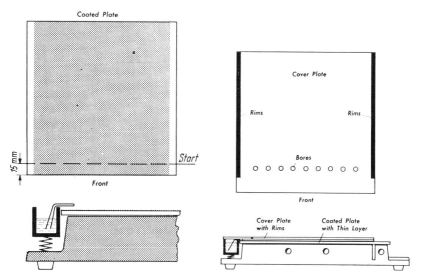

Fig. 23. BN chamber. Coated plate with layer removed on both edges; assembly of coated plate and paper tongue (Desaga GmbH, Heidelberg, Germany).

Fig. 24. BN chamber. Cover plate with rims and bores; assembly of cover plate (Desaga GmbH, Heidelberg, Germany).

As mentioned earlier, the considerable time requirement is a disadvantage of multiple development. The use of the BN chamber (p. 16) permits a stepwise reduction of the developing distance ("iterating" thin-layer chromatography). The coated plate is prepared in the manner already described on p. 16. The samples are applied at a distance of 30 mm from the plate edge. The tank is removed from the stand; the cotton wadding, which is supplied with it, is packed in and is impregnated with about 25 ml solvent. The coated plate is now placed on the cooling block (Fig. 22) with the layer pointing up and its front edge projecting beyond the holder for the solvent tank (Fig. 23). By placing the cover plate with the rings down, the layer is protected. The cover plate is placed in such a manner that the zone between rings (Fig. 24) of 20 mm width projects beyond the heating block (Fig. 25). With this arrangement, a strip of the coating of 20 mm width remains uncovered at the front edge of the layer plate, and the inverted solvent tank is placed on it (Fig. 25). After the first run, the coated plate is removed from the chamber and dried. The layer is separated by a line 3 cm behind the slowest spot, and the plate is again installed. Now, however, it is moved back so far that the slowest spot is located at the point indicated in Fig. 26.

Finally, the tank is placed directly before the separating line (Fig. 26). With the second run completed, the process can be repeated.

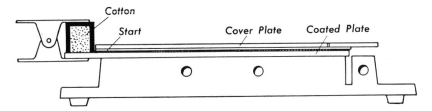

Fig. 25. BN chamber. Assembly for iterating thin-layer chromatography (Desaga GmbH, Heidelberg, Germany).

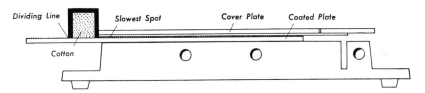

Fig. 26. BN chamber. Assembly for the second run in iterating thin-layer chromatography (Desaga GmbH, Heidelberg, Germany).

E. CONTINUOUS FLOW CHROMATOGRAPHY AND WEDGE-STRIP TECHNIQUE

In the case of small differences in the R_f-values, the development distance must be increased to obtain complete chromatographic separation. This requirement can easily be satisfied for slowly migrating substances by allowing the solvent to flow beyond the zone of the layer. In paper chromatography, this method, known as continuous flow technique, can be realized very simply; chromatography proceeds with the descending method, and the solvent is allowed to drip off from the lower end of the paper.

Mottier [261] was the first to combine thin-layer chromatography with a type of continuous flow method. However, this procedure is limited to the use of solvents with a relatively low volatility. The method of

Brenner and Niederwieser [43] and the BN chamber, which resulted
from it, is very well suited for continuous flow development. Recently,
Ritschard [343] described a modification of the BN chamber.

1. Continuous Flow Development in the BN Chamber

a) *Variant A.*—Assembly takes place as in horizontal thin-layer
chromatography (p. 16). If the solvent front now reaches the end of the
plate, the solvent can be evaporated (possibly with heating). With a
longer flow time, the empty solvent tank is connected with the solvent
reservoir by Teflon® tubing (Fig. 22), and the level in the solvent tank
is established by the glass tube. After opening of the connecting stopcock,
the tank fills to the level of the lower edge of the glass tube in the reservoir (Mariott principle). The quantity of solvent in the tank now remains constant during the entire chromatography (for cooling and
heating, see Fig. 22).

b) *Variant B.*—The procedure is analogous to variant A, but the
application of samples is delayed at least until the solvent front has
reached the end of the separation distance. The adsorbent is then in
equilibrium with the solvent. The substances to be chromatographed are
applied on the wet layer through the holes in the cover plate. This procedure is to be recommended especially for sensitive substances. Variant
B still has other advantages: the unit can be operated continuously for
many days; it is possible to start chromatograms of colored compounds
(possibly with UV-light) in the same unit at any time. Chemical reactions, for example, can be conveniently observed in this manner.

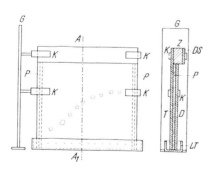

Z—cotton
K—metal clip
P—cardboard or glass strip (5 mm width)
T—coated plate (20 × 20 cm)
D—cover plate (15 × 20 cm)
DS—narrow plate (5 × 20 cm)
LT—tray
G—support base

Fig. 27. Equipment for continuous flow chromatography (from Bancher et al. [7]).

2. Other Continuous Flow Methods

Bennet and Heftmann [22], Truter [426], and Bancher et al. [7] prefer ascending methods. Fig. 27 shows the experimental arrangement used
by Bancher et al. [7].

The solvent is absorbed at the upper end of the layer by means of
clean surgical cotton. The latter is placed between two filter papers and

is pressed firmly on the layer by means of a 5 × 20 cm plate and two steel clamps. A modification of the S-system (p. 15) is used for development. The size of the cover plate selected is such, however, that a layer of about 50 mm width on which the cotton is applied remains exposed. In this manner, the cotton can be changed as often as necessary.

3. Wedge-Strip Technique

In addition to continuous flow chromatography and multiple development, the wedge-strip technique described by Matthias [248] can be used for the separation of substances with similar R_f-values. The use of this method in which the substances migrate in bands in thin-layer chromatography was suggested by Stahl [407]. Following Stahl, Seiler [384] used the wedge-strip technique for the separation of DNS-amino acids (p. 146, Fig. 76).

F. MULTIDIMENSIONAL CHROMATOGRAPHY

As a rule, the separation of complex mixtures, such as amino acids in protein hydrolyzates or amino acids in biological material, for example, is not possible by one-dimensional chromatography. In such cases, it will be of advantage to develop with a second solvent perpendicular to the direction of development of the first solvent. With a suitable selection of the solvent combination, most protein amino acids can be separated, for example (p. 72). In the investigation of complex mixtures (body fluids, organic and plant extracts), a two-dimensional separation is often incomplete. By a skilled combination of several solvents (see "multidimensional chromatography"), however, a good separation can be achieved.

1. Two-Dimensional Chromatography and Separation-Reaction-Separation Technique

In two-dimensional chromatography, the chromatogram is developed with a second solvent perpendicular to the development direction of the first solvent after an intermediate drying step. (The conditions for intermediate drying must be standardized, because the separation in the second dimension is otherwise subject to greater variations.) The second solvent naturally must be selected in such a manner that it resolves those spots which contain several substances. Identification can be made by reference chromatograms. Known mixtures of substances are spotted on the plate edges in both directions. Another possibility is to chromatograph the test solution together with a standard mixture from the same start. The standard solution should contain just enough of each compound so that the corresponding components can just be detectible after the two-dimensional separation. The presence of a certain substance is indicated by an increased intensity of its corresponding spot. The spot pattern of a two-dimensional chromatogram is subject to certain variations (see p. 71). In any case, it must be recommended that every laboratory prepare its own spot patterns. The use of specific or polychromatic reagents (p. 92) is also useful for identification. In cases of doubt, the particular spot must be eluted from the chromatogram. For the purpose of identification, the sample is rechromatographed in another solvent together with the reference substance. Such a procedure, however, is possible only if the substances are present without decomposition after staining or can be recognized without dyes (for example, UV-light). For the non-destructive detection of amino acids, 2, 4-Dinitrofluorobenzene (p. 93) is suitable.

Free amino acids can be chromatographed with the one-dimensional method using solvent A, and if a color reaction is then carried out with the reagent described on p. 92, the amino acids are converted into their dinitrophenyl (DNP) derivatives (see table at back of book). The DNP-amino acids can now be developed in the second dimension with solvent B. This modified SRS (separation-reaction-separation) technique can render valuable service in identification.

In a two-dimensional chromatogram which was developed in both directions with the same solvent, all substances are arranged along the plate diagonal [85] (Fig. 28). Spots which deviate from the diagonal are decomposition products. Thus it can be easily determined whether the substances decompose during chromatography. In the SRS technique an intentional change is produced under the influence of chemical reagents [41, 85, 403].

2. Multidimensional Chromatography

In very complex mixtures of substances, the separation and identification often is not successful with a combination of two solvents. An ex-

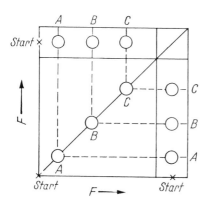

Fig. 28. Two-dimensional development with the same solvent (F) in both directions: detection of decomposition products.

tensive differentiation is possible for example with the use of a third solvent. Multidimensional development was proposed in paper chromatography by Decker [87, 88] and, subsequently, was introduced in TLC by von Arx and Neher [5]. The principle of the method is illustrated in Fig. 29 [5].

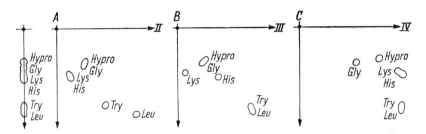

Fig. 29. Principle of multidimensional thin-layer chromatography [5]. Note that each substance appears separately in one of the three two-dimensional chromatograms (from von Arx and Neher [5]).

The sample solution is applied on three thin-layer plates (A, B, and C), and each plate is developed with solvent I in the first dimension. Now each plate is chromatographed in the second dimension with different solvents for each (II for A, III for B, and IV for C). In the resulting three two-dimensional chromatograms, the sequence of amino acids in the first dimension is the same; however, the sequence differs in the second dimension. By a suitable selection of solvents, 52 amino acids can be separated, for example (p. 81).

The interpretation of a chromatogram, which may at first seem complicated due to the large number of spots, is facilitated by allowing a test mixture to run in the second direction after the first run (Fig. 64). In cases of doubt, it is advisable to proceed with the identification in a manner similar to two-dimensional chromatography.

In paper chromatography, Decker [87] proposed the combination of four solvents in six two-dimensional chromatograms. In this case, each substance appears on a total of six independent chromatograms with different solvent combinations.

G. POLYZONAL THIN-LAYER CHROMATOGRAPHY

If solvents consist of several components, they are subject to a partial separation in penetrating a dry adsorbent (p. 54). In polyzonal thin-layer chromatography according to Niederwieser and Brenner [273], substances are separated with the intentional use of this phenomenon. The BN chamber (p. 16) is used for chromatography.

The sample solution is spotted several times at increasing distances from the paper strip (Fig. 30a).

The substances to be separated thus obtain the opportunity to migrate at different times in solvents which have a different composition than the mobile phase in the solvent tank. In this manner, practically the same amount of information is obtained in one run as if the mixture were chromatographed on different plates in different solvents. During chromatography with a two-component solvent, three extreme cases are observed: all substances migrate in the α-zone; they migrate in the β-zone; or they migrate in both zones. (Figs. 30a, b, c). A superposition of these cases is shown in Fig. 31a.

If several substances migrate with the demixing line, it is of advantage to select a solvent with components that do not differ as much in polarity. Frequently, a separation is achieved by changing to polyzonal operation (multicomponent solvents). If the polar component of the solvent is substituted by a mixture of similar polarity, the demixing line is fragmented. Substances which have moved with the demixing line of

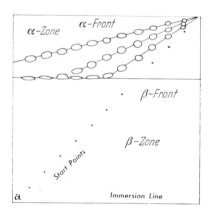

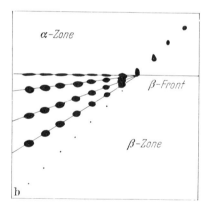

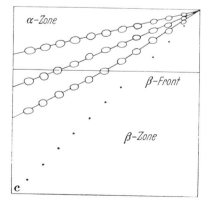

Fig. 30. Polyzonal thin-layer chromatography: The substances are spotted along the plate diagonal [273].

a. The substances migrate in the α-zone.
b. The substances migrate in the β-zone.
c. The substances migrate in both zones.

From Niederwieser and Brenner [273].

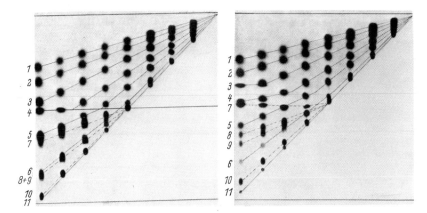

Fig. 31. Chromatograms of mixtures containing 1 µg each of the 2,4-dinitro-phenyl derivatives of n-amylamine (1), n-butylamine (2), n-propylamine (3), ethylamine (4), methylamine (5), tyramine (6), leucine (7), methionine (8), proline (9), hydroxyproline (10), and glutamine (11).

a. Dual-zone development: isopropylether/propionic acid/glacial acetic acid (100:2 v/v).

b. Polyzonal development: Isopropylether/propionic acid/glacial acetic acid/formic acid (100 : 0.66 : 0.66 : 0.66 v/v).

Note that separation with the polyzonal technique is complete. From Nieder-wieser and Brenner [273].

a two-component solvent now can migrate away from each other. An example of this type is shown in Fig. 31b.

The quality of the separation depends upon the following ratio:

$$F = \frac{\text{distance of paper tongue—start}}{\text{distance of paper tongue—}\alpha\text{-front}}$$

In order to test the usefulness of a solvent, the test solution is applied on an oblique start line (Fig. 30a) with a spacing of about 2 cm. The paper tongue should project only 5 mm over the plate, and the development distance of the paper tongue should amount to about 15 cm. It is of advantage to standardize all distances and the position of the start line. The optimum F-ratio determined is important. Mixtures of equimolar quantities of the first members of homologous series generally are more suitable than higher members for the preparation of solvents.

H. GRADIENT DEVELOPMENT

It is often of advantage to change the composition of the solvent continuously during chromatography. This technique, known as gradient elution, is frequently used in column chromatography of protein hydrolysates, for example (e.g. [6]). A further possibility for gradient elution consists of a continuous change of the composition of the *layer* (chromatography on gradient layer).

1. Gradient Elution

The gradient elution technique for thin-layer chromatography was described by Wieland and Determann [442] and by Rybicka [354].

Method of Wieland and Determann [442]: In a glass cylinder (diameter 6 cm and height 30 cm), a chamber of about 1 cm height is formed by a removable filter plate (1) (Fig. 32); the chamber contains a magnetic stirrer (2), and its outer wall is provided with a fused capillary inlet line (3). The overflow (4) is placed 1 cm above the perforated tray. The thin-layer plate (5 × 20 cm) is placed on the perforated tray [filter plate] and is immersed about 1 cm in the solvent. In order to prevent damage to the layer during stirring, a filter paper strip (7) of 1.5 × 5 cm is fastened to the immersed end with the aid of a rubber ring (6). The substances are spotted on the plate at 2.5 cm from the lower edge. For 20 × 20 plates, the dimensions must be modified accordingly. For ion exchange chromatography, the gradient is produced by allowing a second salt solution to flow into the mixing chamber from a burette or from a metering pump through feed line (3) and by mixing with the aid of the magnetic stirrer. The volume is maintained constant by the overflow (4).

Rybicka [354] uses a somewhat simpler device (without maintaining a constant volume), but care must be taken that the start line does not submerge under the solvent at the end.

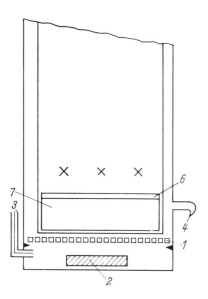

Fig. 32. Separation chamber for gradient elution according to Wieland (see text for explanation). From Wieland et al. [442].

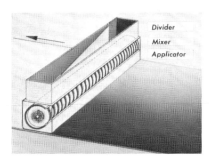

Fig. 33. Semi-schematic representation of the GM-applicator during preparation of a gradient layer. (Desaga GmbH, Heidelberg, Germany.) According to E. Stahl [404].

Recently, Niederwieser and Honegger [476] have been very intensively engaged in the study of the gradient technique. The equipment developed by these authors has proved useful for the analysis of various substances, e.g. lipids (Niederwieser [474]), nucleic acids and related derivatives (Pataki and Niederwieser [477]), and amino acids (Niederwieser [475]). A system for gradient elution TLC is commercially available from Desaga GmbH.

Fig. 34. Front view of the assembled GM-applicator (Desaga GmbH, Heidelberg, Germany). According to E. Stahl [404].

Fig. 35. Separate parts of the GM-applicator. Left: Attachable diagonal divider; center: applicator with gear drive; right: mixing shaft with 30 chamber disks (Desaga GmbH, Heidelberg, Germany). According to E. Stahl [404].

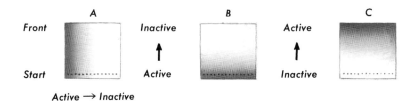

Fig. 36. Different possibilities of application of the gradient layer. The arrows indicate the path of the gradient. According to E. Stahl [404].

2. Chromatography on Gradient Layers

Equipment for the preparation of gradient layers has only recently become known. Berger et al. [26] modified the Stahl applicator (p. 4) for the preparation of multiple layers (p. 97). Later, Stahl [404] designed the GM-applicator (G = gradient, M = mixer), the operation of which is illustrated in Fig. 33.

With the use of the GM-applicator (Figs. 34, 35), it is possible to obtain a continuous transition between two adsorbent properties. The different applicabilities of a gradient layer are shown in Fig. 36.

Variant A is particularly interesting. In this case, numerous separating layers with different properties are present side by side (the most active is on the left and the most inactive on the right). The mixing ratio which produces an optimum separation in a certain separation problem can be determined in one run. In Variants B and C, the activity of the layer changes from start to the front (Fig. 36). The applications for gradient layers are very diverse and cannot yet be predicted. Their use in combination with gradient elution has opened new paths in thin-layer chromatography.

I. PREPARATIVE THIN-LAYER CHROMATOGRAPHY

Valuable information on the composition of an unknown mixture is obtained in analytical thin-layer chromatography. Often it is of advantage and desirable to obtain the individual components in larger quantities, for example, to determine the structure of unknown compounds. On one hand, "classical" column chromatography can be combined with thin-layer chromatography but, on the other hand, a new version of column chromatography and chromatography on thick layers are continuously increasing in importance.

1. Relations to Column Chromatography

Numerous experiments of Stahl [402] have demonstrated that a transition from analytical thin-layer chromatography to preparative column chromatographic separation is possible in many cases. A flow sheet for the control of column chromatography has been presented by Stahl [402]. Duncan [101] and Palmer [291] were intensely occupied with the direct application of thin-layer chromatography to the "classical" column. According to Duncan [101], it is possible to verify on a thin-layer plate whether an adsorption medium (solvent and adsorbent) is also suitable for the column-chromatographic separation of a mixture. The possibility of separation is given by r:

$$r = \frac{R_{fA}}{R_{fB} + 0.1 \cdot R_{fA}} > 1$$

(R_{fA} = R_f-value of the faster moving substance; R_{fB} = R_f-value of the slower moving substance.)

If r = 1, no separation is obtained on the column. The usefulness of r has been demonstrated with several examples [101]. Palmer [291] found that r = 1.3-1.5 is necessary for a complete column-chromatographic separation. The layer preparation with methyl alcohol-water (1 : 1), used by Duncan [101], is rejected by Palmer [291] because of the reduced separating efficiency of such layers. The Canadian author [291] furthermore pointed out that the purity of two separated substances often leaves something to be desired even if r = 1.3-1.5.

A nearly complete preparative analogue to horizontal thin-layer chromatography in the BN chamber (p. 16) was described by Dahn and Fuchs [78]. This method permits a direct transfer of thin-layer chromatography to the horizontal column.

Seamless cellophane tubes, packed with dry adsorbent (e.g. silica gel of < 0.08 mm particle size), served for the chromatography. The weight ratio between adsorbent and substance amounts to 1000 : 1 (silica gel) and 5000 : 1 (cellulose). Development takes place horizontally. For isolation of substances, the column is sliced and the zones are eluted with suitable solvents. Since the R_f-values in the BN chamber, on one hand, and in the Fuchs column, on the other, agree to some extent (Table 1), it is also possible to detect colorless compounds which are invisible in UV-light on the basis of the R_f-values. Admittedly, this is only true with the use of solvents of low volatility. Another possibility for the detection of colorless substances consists of slitting the cellophane tube on its length, introducing a filter paper strip into the slit which be-

Table 1. Comparison of R_f-values on thin-layer plates[1] (Silica Gel G) with R_f-values on the horizontal column according to Dahn and Fuchs [78] (silica gel: particle size < 0.08 mm).

Solvent	Substances	R_f-values	
		Plate	Column
Benzene	Indophenol	0.11	0.13
	Sudan Red G	0.22	0.24
	Butter yellow	0.58	0.59
Ethylacohol/water	Arginine	0.01	0.05
(7 : 3 v/v)	β-alanine	0.39	0.45
	Alanine	0.56	0.57

[1] Horizontal technique, covered plate (see text, p. 15).

comes impregnated with the solvent (and substance). Subsequently, a suitable color reaction is performed on the paper strip. In this manner, the separation of arginine, alanine, and β-alanine (Table 2) can be followed, for example.

Table 2. Control of fractions in the Dahn-Fuchs column by thin-layer chromatography (horizontal technique, covered plate). According to Dahn and Fuchs [78].

Length of column section from start, cm	Zone fulcrum (cm from start)	Content
0.0— 1.6	1.0	Arginine
1.6— 2.8		Small amt. of arginine
2.8— 8.6		Blank
8.6—10.4	9.0	β-alanine
10.4—11.2		Nearly blank
11.2—12.4	11.5	Alanine

2. Chromatography on Thick Layers

Preparative chromatography on thick layers has been described by Ritter and Meyer [345] and by Honegger [164]. Subsequently, numerous authors were occupied with the same technique [166, 150, 291, 208, 79, 22, 406].

For preparative chromatography, loose non-adhering layers of 2-3 mm thickness were already used by Mottier [262] in 1955 and later by Cerny et al. [61] and by Davidek [82]. Ritter and Meyer [345], Palmer [291], and Dauvillier [79] reported that thicker layers frequently tend to form cracks. The separation also is reported to be impaired with increasing layer thickness [345, 406]. Palmer [291] used methanol-water (1 : 1 to 9 : 1) in place of water for the preparation of layers of 3 mm thickness and found that, although no further cracks formed, the resolving power of such layers was poor. The author [291] attributed this finding to an *irreversible methylation of the silanol groups*. On the other hand, if 2 parts dimethoxyethane-water (1 : 1 to 9 : 1) and 1 part silica gel are used, layers of 0.5-3 mm result which satisfy all requirements.

For the preparation of layers of 1-5 mm, Honegger gives the following instructions [164]. Adsorbent and water are vigorously shaken for 30 sec. (in an Erlenmeyer flask) and the suspension is filled into the chamber of the Camag unit (Fig. 5). The viscous mass is well spread by tipping the chamber sideways. When the plates are pushed through uniformly (the semi-automatic Camag unit probably is particularly suited for this purpose; Fig. 6), homogeneous layers are produced. Any air bubbles which are present are burst with a needle while the layer is still liquid. The Stahl applicator and other spreading equipment (Figs. 2 and 3) naturally can also be used.

Air-drying of the layer is supplemented by infrared irradiation (layers of 2-5 mm are dried in air for 60 min. and are simultaneously irradiated with IR; with thinner layers, a 5 min. air drying is sufficient). The ratio of adsorbent-water can be seen in Table 3.

Layers of up to 3 mm can be prepared in this manner without difficulty. An addition of 2% calcium sulfate contributes to the prevention of cracks in thicker layers.

Table 3. Preparation of adsorbent layers of different thickness. From: Honegger [164].

Adsorbent	Layer thickness	Adsorbent: water ratio
Silica Gel G	1.0—1.5 mm	1 : 1.7
	2.0—3.0 mm	1 : 1.6
	3.0—4.0 mm[1]	1 : 1.6
	4.0—5.0 mm[1]	1 : 1.57
Aluminum oxide G	1.0—1.5 mm	1 : 1
	4.0—5.0 mm	1 : 0.9

[1] Possibly with addition of 2% calcium sulfate.

Generally, the best separations are observed on layers which were dried in air for 72 hrs. or were heated to 110° for 24 hrs. and, subsequently, were inactivated for 48 hrs. in air [166]. Saturated separation chambers (tank chamber or S-chamber with additional "saturation," p. 14) are to be preferred for chromatography.

For the application of substances in the form of bands, the applicator described by Ritter and Meyer [345] and by Stahl [406] is suitable. A corresponding arrangement, obtainable from the Desaga Company, calls for the substances to be applied with a pipette (Fig. 37). Recently, the Camag Company has marketed a similar device ("chromatocharger").

The applied quantity can be increased considerably if larger plates than the standard format are used. Halpaap [150] was the first to describe the use of plates of 100×20 cm. Similar equipment can be obtained from the Shandon Company. Stahl prefers 40×20 cm plates [406]. Fig. 38 shows a separation chamber for the development of 40×20 cm plates. The "multisandwich" system of the Camag Company is also suitable for preparative TLC.

Detection of substances invisible under UV is usually done best by spraying narrow strips on the plate edges. The zones are then wiped off

Fig. 37. Automatic system. Application of substance in bands for preparative thin-layer chromatography (Desaga GmbH. Heidelberg, Germany).

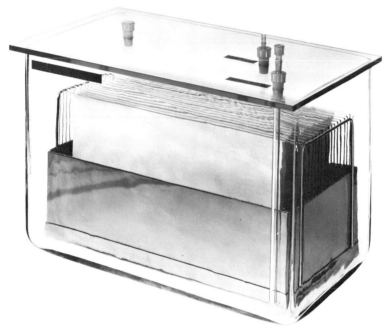

Fig. 38. Separation chamber for the development of several 40 × 20 cm plates (Desaga GmbH, Heidelberg. Germany).

and the substances are eluted. Small "vacuum cleaners"* [345, 261] are suitable to remove the layers; however, care is indicated in the case of oxygen-sensitive substances.

J. THIN-LAYER ELECTROPHORESIS

Thin-layer electrophoresis was described by Honegger [163] and Pastuska [313] at the same time.

As early as 1946, Consden, Gordon and Martin [72] used silica gel for the electrophoretic separation of amino acids and of peptides. However, the method was soon replaced by paper electrophoresis. Attempts were made to find a suitable stationary phase as a substitute for paper. For example, we may note cellulose-acetate films or thin layers of different gels. Undoubtedly, microelectrophoresis on thin layers, which found application in clinical chemistry as well as in biochemistry, can be considered as the direct precursor of present-day, thin-layer electrophoresis.

Honegger [163] and Pastuska [313] made use of silica gel, kieselgur, and alumina as the support and found that thin-layer electrophoresis was equivalent, or even superior, to paper electrophoresis for the separation of amino acids and other substances. The advantages of the thin-layer technique become particularly apparent in the combination of electrophoresis and chromatography [163, 440].

The "fingerprint technique" introduced by Ingram [177], where enzymatic cleavage products are separated by electrochromatography for the comparison of proteins and polypeptides, has recently been applied to thin layers by Wieland [440, 441], Ritschard [343], Fasella [110], Stegemann [414], and by Glaesmer et al. [464].

Instruments for low-voltage electrophoresis were described, for example, by Honegger [163], Pastuska [313], and Nybom [281]. The unit described by Moghissi [255] and Ritschard [343] can also be used at higher voltages. Fig. 39 shows a simple device according to Moghissi [255].

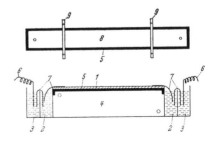

Fig. 39. Equipment for thin-layer electrophoresis. (1) Support plate; (2, 3) reservoirs for buffer solution; (4) cooling chamber; (5) rubber seal; (6) electrodes; (7) asbestos paper; (8) lid; (9) clamps. According to Moghissi [255].

---

* Desaga.

The support plate is directly cooled with water. The cooling chamber is not covered at the top; the plate is placed on the cooling chamber, which is provided with a rubber gasket of 3 mm width on the upper edges. The lid, which also has a rubber seal, is placed over the support plate of identical size and clamped down. The cooling chamber thus is

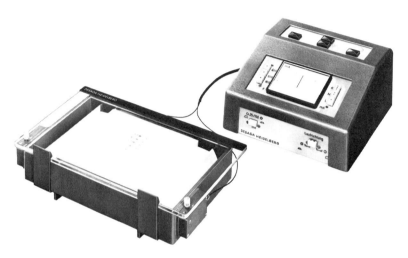

Fig. 40. Thin-layer electrophoresis. Electrophoresis chamber with cooling block and electrical wiring accessory (Desaga GmbH, Heidelberg, Germany).

sealed. At higher voltages, a cooling element must be connected, like that built into the Pherograph.

Fig. 40 shows apparatus of the Desaga Company.

K. Radiochromatography

Thin-layer chromatography is suited for the separation and preparative isolation of radioactive substances. The latter method is particularly important for the determination of the structure of unknown metabolites. With the help of double-labeling, it is possible, for example, to determine the molecular weight on a scale of μg. The tracer technique is also excellent for the quantitative determination of amino acids (p. 61). The qualitative identification of a labeled compound can be obtained, on one hand, with customary chemical reactions; but on the other hand, the radiation can be determined qualitatively and quantitatively by means of suitable instruments [239].

1. Autoradiography

For the purpose of autoradiography, the chromatogram is contacted with a high-speed film and is exposed in the absence of light. With longer exposure periods, it is advisable to store the chromatogram at lower temperature.

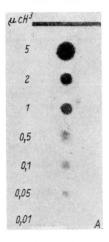

Fig. 41. Detectability of H^3 on thin-layer chromatograms by autoradiography. According to Sheppard and Tsien [338].

The exposure period depends upon the emitter, the radiation intensity, and the sensitivity of the film. The detection limit is in the order of about 0.05 μC for soft radiation. Sheppard and Tsien [338] found that 0.01 to 0.05 μC H^3 could still be detected after an exposure time of one week (Fig. 41).

Fray and Frey [126] reported that the detectable C^{14}-activity amounts to about 0.05 μC with an exposure time of 30 min., but to 0.00005 μC after an exposure of 268 hrs. These authors [126] used high-speed panchromatic film corresponding to 1250 ASA (32° DIN). Phenidon was used as the developer. Fig. 42 shows the relation of the detectability to the exposure time.

A semiquantitative evaluation can be made by simultaneous chromatography and exposure of reference solutions of known activity. Photography of the autoradiograms and subsequent densitometry even permit a more accurate quantitative evaluation [388].

2. Evaluation with Counter Tube

Schildknecht and Volkert [363] divide a chromatogram of 1 cm width into intervals of 0.5 to 1 cm in length, scrape off the layer, and determine the activity with a methane flow counter. A device for the automatic transport of the plate with simultaneous measurement of the activity has

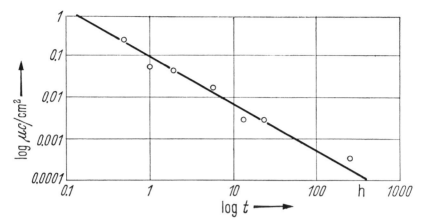

Fig. 42. Dependence of the detectability (C[14]) upon the exposure time [126] in autoradiography. According to Fray and Frey [126].

been described by Schulze and Wenzel [374]. A commercial thin-layer scanner is shown in Fig. 43.

The activity distribution is recorded fully automatically without altering the layer and the substances in any manner. The counting efficiencies are about 20-30% for C[14] and S[35], 50% for P[32], and about 0.6% for H[3].

3. Evaluation with Scintillation Counter

Although H[3]-labeled compounds generally are inexpensive and their toxicity is low, the low counting efficiency is often considered a disadvantage for direct evaluation. And if the ratio between labeled and unlabeled substances is unfavorable, it is advisable to perform the quantitative determination with a scintillation counter. The activity distribution can thus be determined with a thin-layer scanner (Fig. 43). The reagents used for the chemical detection of the substances should not influence the counting rate. Baxter and Senonen [18] used trinitrobenzene-1-sulfonic acid as a spray reagent for the chemical localization of C[14]-amino acids. With this method, the radioactivity is extensively preserved (96-100%). In contrast, if Ninhydrin is used for the detection of 1-C[14]-glutaminic acid, approximately 24-45% of the activity is lost. The method described on p. 92 probably is also suited for the detection of labeled amino acids.

The active zones found are scraped from the plate and are transferred into a Try-Carb counting chamber. Counting can take place without elution: 15 ml of the "cocktail" (5 g PPO + 0.3 g dimethyl-POPOP in 1 l toluene homogenized with 4% CabOSil)[18] are placed into the counting chamber and the C[14]-activity is measured. [PPO = 2,5 diphenyloxazole;

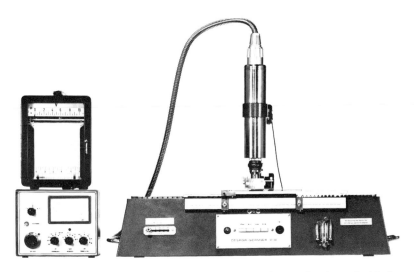

Fig. 43. Automatic instrument for the quantitative evaluation of thin-layer radiochromatograms (Desaga GmbH, Heidelberg, Germany, and Labor Prof. Berthold, Wildbad, Germany) with recorder and scaler.

POPOP = p-bis-2-(4-methyl-5-phenyloxazolyl-benzene.] Conditioning to establish equilibrium is necessary for 30 min. before the H^3-activity is determined. C^{14}- or H^3-toluene serves as the internal standard. (Supplier of the Tri-Carb counting chamber, PPO, and POPOP is Packard Instrument Co., La Grange, Ill.)

The conversion of an unknown labeled compound with a reagent labeled with another isotope permits a molecular weight determination on the scale of μg [437]. For example, with a known specific activity of a tritiated substance, the weight of the substance can be calculated by measuring the absolute activity values. If this unknown tritiated compound is now reacted with a C^{14}-labeled reagent of known molar-specific activity, the molecular weight can be calculated on the basis of the two activities (H^3 and C^{14}) of the derivatives. The number of groups converted per mole must be known, however. Since the calculation is based not on an absolute measurement but on a determination of the ratio of H^3/C^{14}, it is not even necessary to isolate the compounds. The activity measurement can take place either with a thin-layer scanner or with a scintillation counter [436a]. The molecular weight determination of leucine requires about 20 μg H^3-leucine. The sample is converted with 1,1-C^{14}-acetanhydride and, after iterative chromatography on paper and Silica Gel G layers, the activity is measured [437]. A molecular weight of 128.3 $\pm$ 2.6% has been found for leucine (calculated 131) [437].

L. INSTRUCTIONS FOR THE PREPARATION OF THIN-LAYER CHROMATOGRAMS

The most important techniques in thin-layer chromatography were discussed in the preceding sections. In the following, a summarizing instruction for the preparation of chromatograms will be presented. The exact preparation of a layer for a certain separation problem will be given in each case with the discussion of a particular separation. The most important variables of influence on the migration rate of a substance are described on p. 39. The following working procedure takes these factors into account and has proved useful in recent years [47].

The adsorbent used during each series of experiments will be obtained from one and the same lot. The available glass plates are coated with a suitable spreader (p. 4) and are *air-dried overnight*. As a rule, layers of 0.25 mm thickness are used for thin-layer chromatography. The substances are spotted at a distance of 15 mm from the lower plate edge. With one-dimensional development, the substances can be placed at a distance of 8 mm from each other. If the volume utilized amounts to more than 2 μl, the liquid must be applied on the layer in portions with intermediate drying.

The distance from the side edge of the plate amounts to a minimum of 1.5 cm for one-dimensional, as well as two-dimensional, development.

Only freshly prepared solvents of well-defined quality and maximum purity are used for chromatography.

If development takes place in a tank (pp. 13, 14), the chamber atmosphere—aside from a few exceptions (p. 81)—must be saturated with solvent vapors. The chamber, which has been lined *completely* with filter paper, is filled with 100-120 ml solvent and is shaken vigorously before the plates are introduced. The chamber may not be opened during chromatography. During development, the temperature must be kept constant; it is permissible to work at higher temperature with viscous solvents (e.g. in phenol/water).

It is indispensable that the development distance and the distance between start and solvent surface be kept constant; this is made possible by the conditions described in the above. If the development distance is limited by a separation line in the layer, the plate must be immediately removed from the chamber when the solvent has reached this line!

On the subject of chromatographs developed in the S-chamber or the BN chamber, see pp. 14 ff.

It may be advisable in certain cases to maintain a constant *atmospheric humidity* during the preparation and chromatography (p. 40).

Chapter 3

Evaluation of the Chromatograms

A central problem of thin-layer chromatography is the characterization and identification of chromatographed compounds. The developed chromatogram is examined in short- and long-wave UV-light and is subsequently sprayed with a suitable color reagent. For the purpose of documentation, a photocopy, photograph, or drawing on parchment paper is suitable; the layer can also be removed from the plate and stored after treatment with Neatan®, a synthetic dispersion offered by Merck AG which permits removal of the layer from the plate. (Compare Barrolier [11].)

In order to designate the position of a substance in the chromatogram, the R_f-value is used:

$$R_f = \frac{\text{migration distance of the substance}}{\text{distance of the front from start}}$$

The relative R_f-value* can also be used instead of the R_f-value:

$$R_b = \frac{\text{migration distance of reference substance i}}{\text{migration distance of reference substance b}}$$

R_f (R_b)-values are valuable constants under defined operating conditions.

In addition to identification, the quantitative determination is of considerable practical importance.

---

* For a definition of the R_f-values with solvent demixing, see p. 54.

A. R_f-VALUES IN THIN-LAYER CHROMATOGRAPHY

1. Reproducibility of R_f-Values, Factors of Influence

As has already been mentioned, the R_f-values are reproducible only under strictly standardized conditions. As in the case of paper chromatography, they depend upon a number of factors [47, 296, 387].

a) *Activity of the Layer.*—The activity of the layer depends upon the nature and quality of the adsorbent, the activation conditions, and the atmospheric humidity. First of all, the different adsorbent types must be distinguished (silica gel, alumina, cellulose, etc.). Even with a limitation of a single adsorbent type, the quality of a given product frequently varies from one manufacturer to another. Occasionally, differences in quality can even be observed in different adsorbent lots from the same manufacturer. The R_f-values, as well as the R_b-values, depend upon the quality of the support [330]. Brenner et al. [50, 51] have proposed, along the line of a working hypothesis, that the quality variations of an adsorbent be eliminated by the introduction of the "chromatography number" (p. 56). The method described by Galanos and Kapoulas [129], in which two reference substances are used for the identification (p. 46), probably is also suited for this purpose.

The adsorbent is usually applied in aqueous suspension on the plates. The activity of the layer depends upon the period of heating as well as upon the activation temperature [94, 95, 320, 112, 165]. In addition, atmospheric humidity while the plates are stored, while the substances are applied, and while chromatograms are developed plays a substantial role [133, 134]. Kelemen and Pataki [195] investigated the reproducibility of the R_f-values on air-dried and activated layers and found that the former furnish more reproducible R_f-values. These studies were conducted with neglect of constant atmospheric humidity. Consequently, the differences in R_f-values of individual activation methods must be evaluated with caution. It is certain, however, that air-dried layers are less sensitive to atmospheric humidity than activated adsorbents.

Geiss et al. [133, 134] found that the separation of hydrophobic substances in apolar solvents is highly dependent upon atmospheric humidity. It is particularly worthy of note that, when aluminum oxide layers are equilibrated at different atmospheric humidity values (21-75% rel. hum.) and are developed at 70% rel. hum., approximately equal R_f-values are obtained. However, if conditioning is carried out at 91% rel. hum., this plate "brings its climate along" and is no longer leveled down during the relatively short running time; i.e. higher R_f-values result.

Recently, the Desaga Company has marketed a conditioning chamber which is suited for the development of chromatograms at constant atmospheric humidity.

b) *Layer Thickness, Plate Format, and Layer Preparation.*—The influence of layer thickness on the migration rate of a substance was found by Stahl [409] in 1956, i.e. before the method was standardized. An extensive influence is noted particularly below 0.25 mm [409, 348, 382]. Pataki [308] and Honegger [166, 168] found that the R_f-values are also dependent on layer thickness—at least formally—in the range of 0.25-3 mm.

Occasionally, preparation of manual coatings on the plates is proposed (p. 8). However, the manual method has been found inferior to the standard method with regard to reproducibility of R_f-values [297, 307].

In coating of narrow plates (e.g. 5 × 20 cm), the layer thickness frequently is not uniform even with the use of an applicator.

c) *Chamber Saturation.*—The rate of migration of a substance is influenced substantially by the degree of saturation of the tank atmosphere [402, 410, 166, 238, 227]. If the solvent evaporates from the layer during chromatography, the quantity of solvent required to wet the distance between start and front is increased. The period of material transport thus is also increased and higher R_f-values are observed. In 1958, Demole [89] observed that, when solvents consisting of components of different polarity were used, the R_f-values were lower in the center of the plate than at the edges (Fig. 44) (compare also [227]).

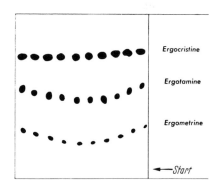

Fig. 44. Edge effect in thin-layer chromatography in unsaturated large-volume separating chambers. According to Stahl [402].

Ergocristine

Ergotamine

Ergometrine

←—Start

Demole [89] attributed this observation to differences in layer thickness; Stahl [402] could demonstrate, however, that the "edge effects" are eliminated by increasing the tank saturation (compare also [190]). In multicomponent solvents, a concentration gradient forms along the chromatogram (p. 53). The slope of this gradient depends, among other things, upon the type of tank and its saturation, and it influences the R_f-value. With a high tank saturation, the gradients are flatter than with low saturation. The formation of the solvent gradient is particularly distinct in the BN chamber (p. 16) or in the S-chamber (Fig. 21).

Honegger [165] investigated four types of saturation: unsaturated tank chamber (KN); saturated tank chamber (K); unsaturated S-chamber (SN); and saturated S-chamber (SK) (Fig. 21). If a mixture of dyes (see indophenol, Sudan red, butter yellow) is chromatographed with benzene or chloroform, then a different sequence of R_f-values can be observed from one type of saturation to another. In benzene, the R_f-values decrease in the following sequence: K-SN-SK-KN; in chloroform, the sequence of decreasing R_f-values is KN-K-SK-SN (Fig. 45). This contradiction can be easily understood when we realize that chloro-

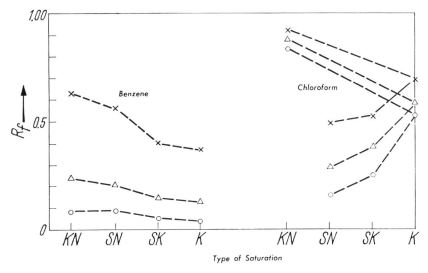

Fig. 45. Influence of chamber saturation in thin-layer chromatography [165]:
×——×   Butter Yellow;
△——△   Sudan Red;
○——○   Indophenol.
KN—unsaturated tank chamber; SN—unsaturated S-chamber; SK—saturated S-chamber; K—saturated tank chamber.

form contains ethyl alcohol even after distillation. In the unsaturated S-chamber, the R_f-values are reduced extensively because a well-defined second front forms due to solvent demixing. The effect of solvent demixing is less distinct in the saturated S-chamber [165, 183]; it disappears in separating systems of large volume, probably due to the counteraction of diffusion from the vapor phase. (For a detailed discussion, see Niederwieser and Honegger [476].) The behavior of the three dyes in the unsaturated S-system with the use of n-butylalcohol-petroleum ether (1 : 9 by vol.) as a solvent is particularly remarkable. All three compounds have the same R_f-value. This behavior undoubtedly is unfavorable but can easily be eliminated if the polar fraction of the solvent is replaced by a mixture of similar polarity (p. 24).

Honegger's experiments [165] have clearly shown that the S-chamber by no means represents a good saturated separation system [183]. The advantages of the covered plate become clearly apparent only with a horizontal arrangement in the polyzonal technique (p. 23). However, since the covered plate has a very small chamber volume, the saturation of the atmosphere plays a subordinate role with regard to reproducibility.

d) *Chromatographing Technique.*—Differences in chromatographic behavior which occur when we change from ascending chromatography (p. 12) to the horizontal technique (p. 15) are also due to a change in the solvent gradient. In single-component solvents, hardly any differences can therefore be observed [306] although the solvent profile may change.

e) *Development Distance and Distance of Starting Point from the Surface of the Solvent.*—A dependence of the R_f-values upon the development distance is occasionally observed. This applies to single-component as well as to multicomponent solvents if the chromatogram is obtained without chamber saturation. With sufficient saturation of the chamber atmosphere, the R_f-values depend little upon the development distance of a single-component solvent (Fig. 46).

With multicomponent solvents, a dependence may still exist, particularly when the individual solvent components differ highly in polarity. In such cases, the R_f-values also depend clearly upon the distance of starting point to solvent surface (immersion line = submersion level). Naturally, the volume of solvent then also influences the migration rate of a substance. In practice, therefore, the ratio of distances

$$\frac{\text{solvent surface—start}}{\text{solvent surface—front}}$$

must be kept constant.

Admittedly, with the appearance of several fronts, a conventional determination of the R_f-value becomes meaningless (p. 53).

f) *Quality of the Solvent.*—The quality of the solvent plays an important role. Even with the most careful control, differences can arise from one case to another. Important sources of error are later changes of the mixture, such as partial evaporation or chemical reaction. Fresh mixtures of pure solvent should always be used for the determination of R_f-values.

g) *Quantity of Substance and Solvent.*—The relative migration rate of the chromatographed substances usually depends upon the quantity applied (e.g. [42, 61, 411]). As shown by Fig. 47, differences are noticeable only with very small and very large quantities. The chromatogram,

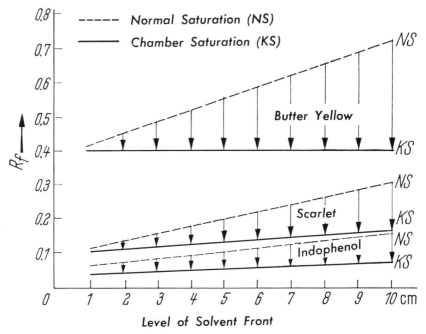

Fig. 46. Dependence of the R_f-values upon the development distance of the solvent (Silica Gel G; benzene). NS—tank chamber without saturation; KS—tank chamber with saturation. According to Stahl [402].

— — — —   Normal saturation (NS)

———————   Chamber saturation (KS)

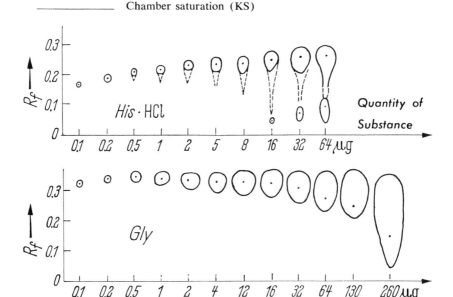

Fig. 47. Dependence of the R_f-values upon the quantity of substance applied [42]. According to Brenner and Niederwieser [42].

and thus the R_f-value, is also influenced by the solvent in which the test substances are applied [195, 312].

h) *Temperature.*—The influence of temperature is a controversial subject in thin-layer chromatography. Jatzkewitz and Mehl [182] as well as Shellard [387] have pointed out that chromatography is influenced little by temperature. Niederwieser [272] investigated the influence of temperature on the R_f-values of amino acids (solvent: phenol-water 75 : 25 g/g) and found that a temperature increase from 18°C to 38°C does not cause significant R_f-changes (Table 4).

Table 4. Temperature dependence of the R_f-values of amino acids on Silica Gel G layers (ascending technique: chamber saturation). According to Niederwieser [272].

| | R_f-value[1] | | |
Amino acid	38°C	18°C	ΔR_f[2]
α-Aminocaprylic acid	0.618	0.606	0.012
β-alanine	0.285	0.287	0.002
Histidine	0.296	0.295	0.001
Hydroxyproline	0.358	0.354	0.004
Methionine sulfone	0.332	0.338	0.006
Phenylaanine	0.555	0.557	0.002
Proline	0.473	0.480	0.007
Tryptophan	0.572	0.562	0.010
Tyrosine	0.431	0.434	0.003

[1] Mean values from 8 single determinations; solvent: phenol/water (75 : 25 w/w).

[2] No deviation is significant (t-test: 95% confidence limit).

Honegger [166] found lower R_f-values at 4°C than at 23°C. However, the number of experiments is too small; these data, consequently, must be viewed with reservations. Stahl [404] chromatographed at different temperatures between −15°C and +20°C and found that some R_f-values increase with increasing temperature, while others are hardly affected.

Geiss et al. [134] obtained chromatograms at constant atmospheric humidity under isothermal conditions at 20, 40, and 60°C and found that the R_f-values at times decrease with increasing temperature. The same authors [134] also studied the influence of temperature at constant relative humidity. Under these conditions, the R_f-values increase with increasing temperature. *It is important that, although the R_f-values are increased at constant relative humidity, the R_f-ratios (R_b-values) do remain constant!*

i) *Contaminants.*—Substances with similar R_f-values have an influence on each other with regard to the distribution between mobile and stationary phase. Displacement phenomena occur in such a case. This admittedly does not explain why the R_f-value of a substance is often lower in a mixture. Shellard [387] is of the opinion that the partial adsorption isotherms deviate from individual adsorption isotherms. An interaction between two substances can occur in the mobile as well as in the stationary phase. If this occurs in the mobile phase, the adsorption should decrease and the R_f-value should increase. If the interaction occurs in the stationary phase, the adsorption should increase. The R_f-value decreases (Shellard [387]).

2. Consequences in Practice

Our considerations show that the R_f-values are reproducible constants only under strictly standardized conditions. Of the factors of influence that have been discussed, (b) to (h) can be maintained constant without difficulties. The activity of the layer (a) can be standardized by a precise control of the drying conditions and atmospheric humidity. It should be possible to eliminate variations due to quality differences in the adsorbent with the use of the chromatography number (p. 56). (For identification of two standard substances, compare Table 6.) We recommend the method described on p. 37, which takes the above mentioned factors into consideration.

In publications, the mean R_f-values, the corresponding standard deviation, and the range of variation of the R_f-values found should be listed at the same time (p. 48). Furthermore, a precise description of the operating conditions (factors a to h) is indispensable!

Finally, we should make reference to the very interesting study of Galanos and Kapoulas [129] who postulate that every variation of R_f-values is linear. For a comparison of R_f-values which are obtained under different conditions, such as in two different laboratories, the following simple equations can be used:

$$a = \frac{R_f{}^I{}_A - R_f{}^I{}_B}{R_f{}^{II}{}_A - R_f{}^{II}{}_B}$$

$$b = R_f{}^I{}_A - a \cdot R_f{}^{II}{}_A \text{ and}$$

$$R_f{}^I{}_i = a \cdot R_f{}^{II}{}_i + b$$

where R_{fA} and R_{fB} are the R_f-values of two compounds of the *same* class of compounds; I and II refer to two independent measurements.

The applicability of the above equations can be explained by the following example [129]: A mixture of 10 amino acids is subjected to paper chromatography. The R_f-values are shown in Table 5. This table also shows the R_f-values resulting from a second independent determination.

Aspartic acid and leucine will be used as reference substances in the calculation:

$$a = \frac{R_f{}^I{}_{Leu} - R_f{}^I{}_{Asp}}{R_f{}^{II}{}_{Leu} - R_f{}^{II}{}_{Asp}} = 1.36 \text{ and}$$

$$b = R_f{}^I{}_{Leu} - 1.36 \cdot 0.57 = -0.055$$

Table 5 contains the $R_f{}^I$-values which were calculated with a and b and which show very good agreement with the directly determined values. Although this method was proposed for paper chromatography, we would like to suggest that it be used in thin-layer chromatography. Preliminary results are shown in Table 6 [306].

Table 5. Paper chromatographic R_f-values of amino acids (solvent: n-butanol-glacial acetic acid-water 12 : 3 : 5 v/v) under different experimental conditions (I and II) [129]. According to Galanos and Kapoulas [129].

Amino acid	$R_f{}^I$		$R_f{}^{II}$
	Det.	Calc.	
Aspartic acid	0.23	—	0.210
Cystine	0.05	0.047	0.075
Glutamic acid	0.28	0.285	0.250
Serine	0.22	0.217	0.200
Threonine	0.26	0.265	0.235
Alanine	0.30	0.305	0.265
Histidine	0.11	0.115	0.118
Valine	0.51	0.509	0.415
Phenylalanine	0.60	0.605	0.485
Luecine	0.720	—	0.570

Table 6. Thin-layer chromatographic R_f-values of PTH-amino acids on Silica Gel G layers (solvent: chloroform-methanol 9 : 1 v/v) under different experimental conditions (I and II) [306].

PTH	$R_f{}^I$		$R_f{}^{II}$
	Det.	Calc.*	
Alanine	0.68	—	0.77
Asparagine	0.23	0.22	0.34
Glutamine	0.28	—	0.40
Glycine	0.56	0.58	0.68
Histidine	0.29	0.28	0.40
Isoleucine	0.77	0.75	0.83
Leucine	0.77	0.76	0.84
Lysine	0.71	0.69	0.78
Methionine	0.75	0.72	0.81
Methionine sulfoxide	0.40	0.43	0.54
Phenylalanine	0.74	0.72	0.81
Proline	0.82	0.81	0.89
Threonine	0.45	0.48	0.58
Tryptophan	0.62	0.62	0.71
Valine	0.74	0.72	0.81

* Reference substances: alanine and glutamine.

3. Comparison with Paper Chromatography

In order to evaluate the instructions given on p. 38, Brenner et al.
[47] chromatographed 42 substances from different classes of compounds on Silica Gel G layers. On the basis of a total of about 850
R_f-values, the standard deviation (^sR_f) was determined, and a mean of
standard deviations ($^{\bar{s}}R_f$) was calculated from the total standard deviations.

An R_f-mean value ($\overline{R}_{fk}$), the standard deviation (^sR_f), and the mean
value of the standard deviations ($^{\bar{s}}R_f$) were determined for each substance k from n R_{fk}-observations ($8 \leqq n \leqq 18$) obtained from n chromatograms on n plates in a suitable solvent; the plates were coated with
silica gel from the same lot.

$$\overline{R}_{fk} = \frac{\Sigma R_{fk}}{n}$$

$$^sR_{fk} = \sqrt{\frac{\Sigma (R_{fk} - \overline{R}_{fk})^2}{n-1}}$$

$$^{\bar{s}}R_f = \frac{\Sigma \, ^sR_{fk}}{n}$$

The calculation was performed for the ascending technique ($n = 60$)
and for the horizontal technique ($n = 20$).

For a comparison, we used 2000 R_f-values for the ascending ($n = 96$)
and 143 R_f-values ($n = 6$) for circular paper chromatography. Table 7

Table 7. Reproducibility of R_f-values in paper and thin-layer chromatography.
According to Brenner et al. [47].

Technique	$^{\bar{s}}R_f$	*No.* ^sR_f
TLC		
Ascending	0.016	60
Horizontal	0.014	20
Paper chromatography		
Ascending	0.018	96
Circular	0.014	6

shows the results. This investigation shows that the variability of the
R_f-values is approximately the same on paper and thin-layer chromatograms.

Brodasky [53] considers the reproducibility of thin-layer chromatography R_f-values in a much more unfavorable light. He investigated
antibiotics on silica gel and cellulose layers and on paper. Unfortunately,
the experiments were conducted under conditions which make the results doubtful. Brodasky chromatographed in large-volume separating
chambers (p. 12) without chamber saturation, although R_f-values can

be considered only with reservations under such conditions, as shown previously on p. 41. The cited author [53] found a superiority of paper chromatography with regard to variability on the same day as well as from day to day.

My own experiments [306] have shown, however, that the daily variability of the R_f-values of amino acids obtained in thin-layer and paper chromatography is approximately the same if the thin-layer chromatography is performed with chamber saturation. We chromatographed four amino acids simultaneously on Whatman No. 1 paper and on Silica Gel G layers. The results are shown in Table 8.

Table 8. Daily variability of amino acid R_f-values on paper and thin-layer chromatograms (solvent: propanol-water 7 : 3, Silica Gel G and Whatman No. 1; ascending TLC with chamber saturation and ascending paper chromatography after conditioning for 16 hrs.) [306].

Amino acid	*Silica Gel G* R_f^1	^sR_f	*Whatman No. 1* R_f^1	^sR_f
Serine	0.32	0.018	0.26	0.014
Proline	0.42	0.022	0.18	0.013
Glycine	0.29	0.008	0.24	0.016
Glutamic acid	0.29	0.007	0.22	0.016

1 Mean values of 6 single determinations.

The mean of the standard deviations amounts to 0.014 for paper and 0.015 for thin-layer chromatography.

B. R_f-VALUE AND CHEMICAL STRUCTURE

1. The Martin Relation

The relation between structure and chromatographic behavior has formed the subject of numerous studies. In 1950, Martin [245] postulated that a linear relation existed between a derived function of the R_f-value and the number of homologous building blocks in the case of paper distribution chromatography. According to Martin we have:

$$R_m = log\,[(1/R_f) - 1] = G_o + mG_x + nG_y + \ldots + G$$

Where:

      G_o is the constant of the elementary group;
      G is the constant of group μ (μ = x,y ...);
      G is the elementary constant.*

---

* The elementary constant is the logarithm of the ratio between mobile and stationary phase and depends not only upon the solvent but also upon the quality of the absorbent.

The linearity demanded by Martin could be confirmed later by a number of authors. (Paper chromatography, see [204, 203, 216, 219, 142, 28, 38, 132, 141, 149]; gas chromatography, see, for example [198].)

a) *Results of Thin-Layer Chromatography.*—In thin-layer chromatography, the relationship between the R_m-value and number of homologous building blocks was investigated by Pataki [295] as well as by Halmekoski and Hannikainen [149]. Pataki [295] chromatographed the homologous series of α-amino acids (glycine to α-aminocaprylic acid) on Silica Gel G layers and confirmed the validity of the Martin relation. The investigated solvents are listed in Table 9 (Fig. 48).

Table 9. Validity of the Martin relation in the thin-layer chromatography of homologous amino acids on Silica Gel G layers (ascending technique, chamber saturation) [295].

Solvent	
Isopropylalcohol/water	70 : 30 v/v
n-Butanol/glacial acetic acid/water	4 : 1 : 1 v/v
Phenol/water	75 : 25 g/g
R^1-OH/glacial acetic acid/water	70 : 2 : 30 v/v
R^1-OH/glacial acetic acid/water	70 : 4 : 30 v/v
R^1-OH/glacial acetic acid/water	70 : 8 : 30 v/v
R^1-OH/glacial acetic acid/water	70 : 16 : 30 v/v

[1] $R = CH_3$; C_2H_5; C_3H_7; iso-C_3H_7.

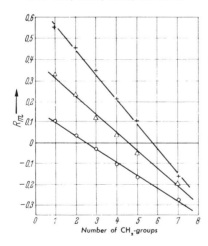

Fig. 48. Relation between the R_m-value and number of CH_2-groups in TLC [295]. Mean values from 18 single determinations (homologous series of α-amino acids).
O   Methanol-water (7 : 3 v/v)
△   Ethanol-water (7 : 3 v/v)
+   n-Propanol-water (7 : 3 v/v)
According to Pataki [295].

Halmekoski and Hannikainen [149] used Silica Gel G and polyamide-Woelm layers for the chromatography of homologous phenols and phenol derivatives; a linear relation could be observed in many cases between the R_m-value and number of CH_2-groups in the para-position.

b) *Deviations from the Martin Relation.*—It is known from paper chromatography that the Martin relation is not applicable in many cases.

This finding caused many authors to introduce partly empirical and partly theoretical, more or less well-founded, corrections into the Martin relation. According to Franc et al. [120, 121, 123, 124, 125], the deviations present in chain-substituted, aromatic compounds can be suppressed in the case of polar stationary phases by the consideration of dipole moments. According to Schauer and Bulirsch [361], the deviations are reportedly eliminated if correction terms are introduced into the Martin equation (p. 49) for the association of organic compounds. Finally, Franc and Jokl [122], who evaluated numerous R_f-tables in literature, found that a linear relation ("log-log relation") exists between the R_m-value and the logarithm of the number of homologous building blocks.

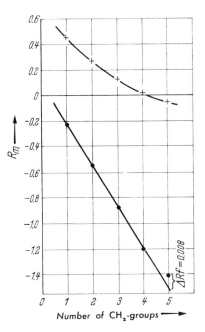

Fig. 49. Relation between R_m-value and number of CH_2 groups (homologous 2,4-dinitrophenylamines) on Silica Gel G layers in ether (·) and benzene (+) as solvents. Note that the Martin relation is valid only in ether. According to Niederwieser [272].

Marked deviations from the Martin relation could also be observed in thin-layer chromatography. Niederwieser [272] chromatographed homologous 2,4-dinitrophenylamines (C_1-C_5) with ether and benzene as solvents and was able to find a linear relation between the R_m-value and number of CH_2-groups *only* in the first case (Fig. 49).

In the case of benzene, a linearity is restored if the R_m-values are plotted against the logarithm of the number of CH_2-groups (Fig. 50, "log-log relation").

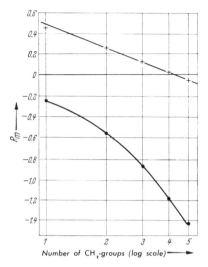

Fig. 50. Relation between R_m-value and number of CH_2-groups (compare legend of Fig. 49). Note that a linear relation exists between R_m-value and log number of CH_2-groups in benzene ("log-log relation"). According to Niederwieser [272].

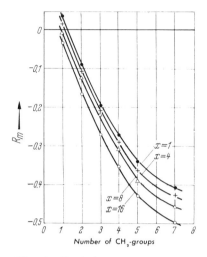

Fig. 51. Deviation from the Martin relation (homologous α-amino acids). Solvent: methylalcohol-water-34% ammonia (70 : 30 : x). Ascending chromatography on Silica Gel G layers. According to Pataki [295].

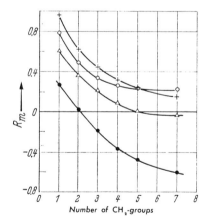

Fig. 52. Deviation from the Martin relation (homologous 2,4-dinitrophenyl-amino acids). Solvents:

+   Chloroform-ethylacetate-glacial acetic acid (80 : 20 : 1 v/v)
o   Chloroform-methylalcohol-glacial acetic acid (95 : 5 : 1 v/v)
△   Toluene-glacial acetic acid (9 : 1 v/v)
●   Ether-glacial acetic acid (95 : 5 v/v)

Horizontal development in the BN chamber on Silica Gel G. According to Niederwieser [272].

Niederwieser [272] discussed the possibility that adsorption chromatography is involved in benzene and that the validity of the Martin relation possibly cannot justifiably be assumed.

Hromatka and Aue [171] found a linear relation between log R_f-value and number of C-atoms. Prof. Hromatka [172] was kind enough to inform us that the "log-log relation" shown in Fig. 50 can also be taken into consideration.

Niederwieser [272] and Pataki [295] also observed cases in which neither the Martin relation nor the "log-log relation" is valid (Figs. 51 and 52).

c) *The Martin Relation in the Case of Chromatographic Solvent Demixing.*—A constant ratio between mobile and stationary phase (expressed by G in the equation on p. 49) is assumed in the Martin relation. As a matter of fact, the solvent distribution is not constant along the development distance [234, 56, 57, 137, 449, 246] and depends upon the chromatographing technique [56, 57, 137, 449].

Giddings et al. [137] determined the solvent distribution ("solvent profile") and found that the "profiles" can be reduced to a characteristic "elementary profile" for a given stationary phase and solvent with a given chromatography technique. We were able to confirm qualitatively that similar "solvent profiles" existed in thin-layer chromatography [272, 295]. Fig. 53 shows the solvent distribution in ascending TLC [272, 295]; similar profiles were also obtained with the horizontal technique [272]. According to more recent studies of Niederwieser [475], the curve at z = 16.5 does not intersect the abscissa, but tends to parallel the abscissa asymptotically.

Fig. 53. Solvent distribution on thin-layer chromatograms. "Profile" as a function of the distance (z) between start line and point of observation (Silica Gel G); ascending technique with chamber saturation; n-butanol/glacial acetic acid/water 4 : 1 : 1 v/v). According to Niederwieser [272] and Pataki [295].

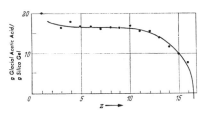

It is important that the phase ratio changes during chromatography. Giddings et al. [137] have shown that the observed R_f-value deviates by about 15% from the "theoretical" R_f-value.

Otherwise, the profile has an approximately horizontal course for average R_f-values (0.15-0.7). If a further advance of the solvent front is prevented on a silica gel layer by making a dividing line, the solvent flow does not stop immediately. After 24 hours, the profile levels out considerably and all R_f-values of the amino acids increase by a factor of 1.12 [42, 272, 295]. If the observed R_f-values are corrected with the empirical factor of 1.12, values are obtained which furnish the "theoretical" R_f-values in good approximation. This deviation between observed and "theoretical" R_f-values, however, has hardly an influence on the linearity of the Martin relation in the R_f-range of 0.1-0.7, as demonstrated by Pataki [295].

We now turn back to the deviations from the Martin relation shown in Figs. 51 and 52. In order to understand this discrepancy, the following observation of Niederwieser [272] is of importance: solvents consisting of several components are subject to a partial demixing as they penetrate a dry adsorbent and, as a consequence, several solvent fronts are observed. A solvent of n-components, as a rule, forms n-zones which are separated by (n-1)-fronts (in polyzonal thin-layer chromatography, this phenomenon is intentionally exploited; p. 23). Fig. 54 shows a simple method for the determination of the individual fronts [272, 273].

Sharp solvent fronts are observed especially in the BN chamber (p. 16).

Thus, it is incorrect to refer the R_f-value to the α-front (Fig. 54) if the substance migrates in the β-zone! In such cases, we determine the following:

$$^\beta R_f = \frac{\text{distance: center of spot-}\beta\text{-front}}{\text{distance: start-}\beta\text{-front}}$$

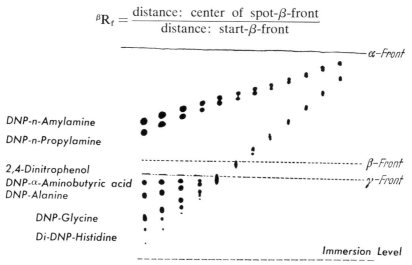

Fig. 54. Solvent demixing: detection of the demixing line. The substances are spotted along the plate diagonal (Silica Gel G; horizontal chromatography in the BN chamber; solvent: chloroform/methylalcohol/glacial acetic acid 95 : 5 : 1 v/v). The α-, β- and γ-fronts are directly visible in UV-light.

Figs. 55 and 56 show that the linearity between the R_m-value and the number of CH_2-groups is restored if the $^\alpha R_f$-values plotted in Figs. 51 and 52 are referred to the "correct" front.

The position of the α-, β-, γ- etc. . . . ω-front is expressed by the retardation factor k; for example:

$$k = \frac{B + s_0}{A + s_0}$$

In the above, B is the distance between start and β-front (Fig. 54), A is the distance between start and α-front (Fig. 54), and s_0 is the distance between immersion line and start.

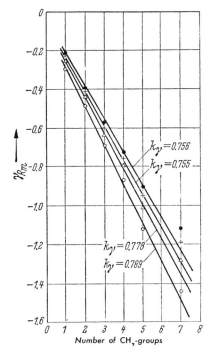

Fig. 55. Validity of the Martin relation after transformation of the $^{\alpha}R_f$-values (Fig. 51) into the $^{\gamma}R_f$-values. According to Pataki [295].

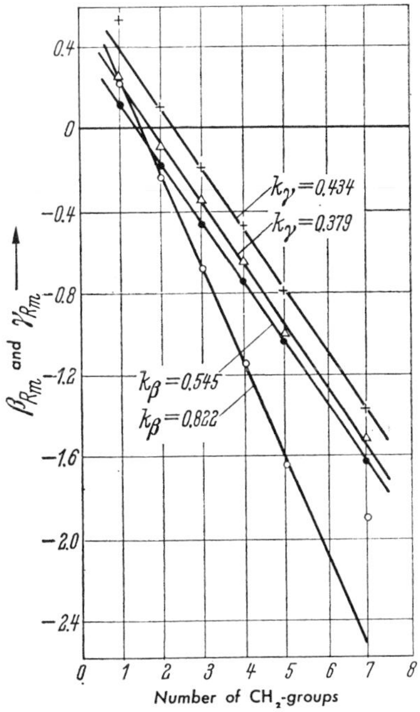

Fig. 56. Validity of the Martin relation after transformation of the $^{\alpha}R_f$-values (Fig. 52) into the $^{\beta,\gamma}R_f$-values, According to Niederwieser [272].

Niederwieser [272] was able to demonstrate that, in the case of large k_β, k_γ, . . . -values (>0.85), the observed R_f-values can be used directly with the application of the Martin relation. The error is within the range of the R_f-determination.

The solvent used for the verification of linearity in Fig. 48 is also subject to demixing. For example, $k_\beta = 0.95$ was found for methanol/water (7 : 3 v/v) [295]. As indicated above, the effects of solvent demixing can be neglected with such a high k_β-value.

2. Applications of the Martin Relation

In a systematic evaluation of the Martin equation (p. 49), it is possible to determine empirical constants. This knowledge permits the calculation of R_f-values, the determination of configurations, or the calculation of the number of functional groups.

It must be taken into consideration that the contribution of a group to the R_m-value of the total molecule may be dependent upon the position of this group in the molecule and upon the rest of the molecule. This problem was given particularly detailed study by Liebnitz et al. [221], Behrens et al. [21], Schauer and Bulirsch [361], as well as by Green et al. [142, 143, 241, 242]. Green et al. showed that consideration must be given not only to "normal" group constants but also to constitutional conditions and even to atomic G_μ-parameters. Finally, we found in thin-layer chromatography [303] that the G_μ-parameters deviate from each other, at least formally, in a series of free and carbobenzoxyamino acids.

All constants in the Martin equation (p. 49) depend upon the adsorbent and solvent used. A change in the system is connected with a change of the R_m-values ($^\Delta R_m$). Macek [236, 237] used this method to determine the number of free α-hydroxyl groups in a series of veratrum alkaloids. Reichl [336] was able to calculate the number of carboxyl groups in organic acids from the $^\Delta R_m$-values. Finally, Macek [235] demonstrated that the $^\Delta R_m$-value differs in the reserpoids of the normal and the iso-series (identification of the configuration). This $^\Delta R_m$-method was expanded in paper chromatography particularly by the studies of Bush [58]. The first beginnings for the use of the $^\Delta R_m$-principle in thin-layer chromatography have already been made [148, 226, 256].

If the R_m-values of two substances (i and st) are determined in the same chromatographic system* under identical conditions, it can be shown that the difference:

$$(R_m)_i - (R_m)_{st} = (R_k)_{i/st}$$

is independent of the quality of the adsorbent [50, 51]. Brenner et al. [50] have suggested the term "chromatography number" for $R_{ki/st}$. The application of the chromatography number should help in eliminating R_f-variations resulting from quality fluctuations of the adsorbent. (The R_c-values defined by Decker [86] probably would also be suitable for this purpose.) It is a prerequisite that all factors of influence (p. 39) are standardized and that the Martin relation is applicable in the particular chromatographic system involved.

C. QUANTITATIVE THIN-LAYER CHROMATOGRAPHY

A quantitative evaluation of the thin-layer chromatogram can be made directly on the layer (*in situ*) or *after elution*.

The test substances must be spotted as carefully as possible for any quantitative determination. Suitable instruments for their application

---

* Under standardized working conditions, a chromatographic system is the stationary phase (adsorbent) and the solvent.

were discussed earlier (p. 12). Every operation with a pipette is subject to certain error. This error can be considerably reduced by increasing the volume applied. If the samples are spotted, the volume naturally is limited, for if the quantities applied are too large, the separation may be impaired under certain conditions. If the method selected for evaluation permits a band application, thicker layers are preferable on which large volumes can be applied without difficulty. For example, if 100, μl or more of a solution are applied on a layer of 1 mm thickness with the use of a blood sugar pipette, the error due to the pipette is negligibly small [36]. In general, the application error can also be reduced with a smaller volume by introducing an "internal standard" as in gas chromatography.

1. In Situ Evaluation

The concentration of a sample can be determined semi-quantitatively, as well as quantitatively, by a comparison of the spot area and/or spot intensity with those of standard substances.

For a semi-quantitative determination, different dilutions of the test and reference solution are spotted alternately on the same plate; after development of the chromatogram and coloring, the spot areas and intensity of individual spots are compared. A variant which is frequently used is the method of maximum dilution. In this case, we apply dilutions of a sample solution as well as the dilution of the standard quantity which can just be made visible. The concentration of a just barely visible spot of the test solution corresponds to the limit of detection of the standard.

In the investigation of biological material, it is advisable to chromatograph the sample and standard together from the same starting point. For example, the sample and a 1 : 1 dilution of it are alternately applied on the same plate. Different standard quantities are then applied on the points loaded with the dilution after an intermediate drying stage. After development and coloring, those spots which coincide in intensity are selected from the sample and from the dilution + known standard quantity. If the unknown quantity of substance in the dilute solution is designated by x and that in the undiluted solution by 2x while y designates the standard quantity, then x + y = 2x and x = y, i.e. the concentration of the dilute solution corresponds to the added quantity of standard.

According to Purdy and Truter [327, 328, 329], a linear relation exists between the square root of the spot area and the logarithm of the applied quantity (Fig. 57).

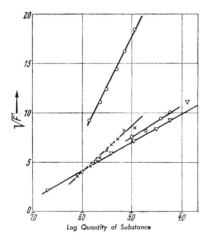

Fig. 57. Relation between the logarithm of the applied quantity of substance and the square root from the
spot area in thin-layer chromatography
of different substances. According to
Purdy and Truter [328].

If the sample A is chromatographed together with its dilution B
and the standard S on the same plate, the following equations are applicable [327, 328]:

$$\sqrt{F_A} = m \cdot \log G_A + c$$
$$\sqrt{F_B} = m \cdot \log G_A \cdot v + c$$
$$\sqrt{F_S} = m \cdot \log G_S + c$$

where:
   F_A is the spot area of sample solution;
   F_B is the spot area of dilute sample solution;
   F_s is the spot area of standard solution;
   G_A is the quantity of substance in A;
   G_S is the quantity of substance in S;
   v is the dilution factor v $(A \rightarrow B)$;
   m and c are constants.

After eliminating m and c [327, 328]:

$$\log G_A = \log G_S + \frac{\sqrt{F_A} - \sqrt{F_S}}{\sqrt{F_B} - \sqrt{F_A}} \cdot \log v$$

The above equation* permits a calculation of the quantity of substance. It is a prerequisite that the samples are applied in equal volumes
on plates of identical thickness.

The determination of the spot area can also be carried out by planimetry or by counting on millimeter graph paper. During planimetry the

---

* Another possibility is represented by the use of the so-called G_r-method
[327, 328].

spot should be evaluated at least 5 times [290, 304]. Table 10 shows the accuracy of the method in the determination of creatinine [304].

Table 10. Accuracy of a quantitative determination according to Purdy and Truter. Quantitative determination of creatinine (quantity of substance: 5 μg). According to Pataki and Keller [310].

Sample	Method I[1]	Method II[2]
1	6.0	5.2
2	4.4	4.0
3	5.1	5.1
4	4.3	5.4
5	5.4	4.7
6	4.5	4.5
7	5.9	4.9
8	5.5	5.3
9	5.4	4.0
Mean	5.2	4.8
Standard deviation	0.64	0.53
Relative standard deviation	12.2%	11.0%

[1] Planimeter circumscribes spot 5 times.

[2] Counting on mm graph paper.

A great advantage of the algebraic method according to Purdy and Truter [327, 328] consists of the fact that no special instruments are necessary.

A direct photometry requires much more equipment. Zöllner et al. [456] have demonstrated that it is possible to evaluate thin-layer plates by means of a densitometer designed for paper chromatography. Squibb [400, 401] made use of plastic disks or glass plates coated with Silica Gel G for the separation of amino acids; the former can be cut into strips and can be evaluated like paper strips.

Hara et al. [152] developed an evaluation instrument which is also suitable for the densitometry of two-dimensional thin-layer chromatograms. The Japanese authors found, among other things, that the reproducibility of a quantitative determination depends highly upon the color development; the uniformity of the layer is reported to be of only subordinate importance; an approximately linear relation exists between the applied quantity and the integrated area [152].

Frodyma et al. [127, 128, 128a] used a Beckmann-DK-2 spectrophotometer for the evaluation of the remission and found that amino acids can be determined with an accuracy of 5-12% [128a].

Fig. 58 shows a densitometer of Joyce, Loebel & Company (Newcastle, England).

Seiler et al. [383, 385] as well as Pataki and Strasky [478] and Connors and Boack [71] reported on a direct fluorimetric evaluation of thin-layer chromatograms. Seiler et al. [383, 480] used a Zeiss spectrophotometer PMQ II together with a mercury lamp ST 41 and the chromatogram attachment CA 2. The best reproducibility of results is obtained if the plates are illuminated from the uncoated side. The plate advance (8 mm/min.) is automatic, and the measured total fluorescence is recorded automatically. Provided the concentrations are not too high, the curve areas are directly proportional to the quantities of substance [383]. An automatic system for direct fluorometry is available from Camag. Table 11 shows the limits of detection, excitation, and fluorescence maxima of some compounds.

Recently, Klaus [200] has been intensively occupied with a direct photometric evaluation of thin-layer chromatograms.

Fig. 58. Densitometer of Joyce, Lobel & Company, Newcastle, England.

Table 11. Limits of detection, excitation, and fluorescence maxima of the formaldehyde condensation products of some compounds. According to Seiler et al. [385].

Substance	Limit of detection μg	E_{max} $m\mu$	F_{max} $m\mu$
Tyramine	0.1	390	500
Tyrosine	0.05	391	500
3-aminotyrosine	0.05	388	510
3,4-dihydroxyphenylalanine	0.05	388	510
Phenylalanine	10.0	—	—
Tryptamine	0.005	388	504
5-hydroxytryptamine	0.05	393	520
Tryptophan	0.005	390	504

2. Evaluation after Elution

After a suitable color reaction (or UV visualization), the chromatographed substances can be eluted from the layer. Barrolier [11, 13] uses the combination of solvents shown in Fig. 59 [42] for the chromatography of free amino acids and sprays the carefully dried plate with Ninhydrin-Cd Reagent (p. 90).

Color reagent is applied on the horizontally placed plate. A sufficient quantity is sprayed until the layer is fully moistened, taking care that the layer does not "float." Approximately 50 ml reagent are used for a 20 × 20 cm plate. The color development takes place at room temperature (20-24 hrs., protected from direct light). As a result of the large excess of reagent, no relation between the forming quantity of color and quantity of reagent exists. The amino acid spots are circumscribed with a fine needle and are subsequently moistened with a small amount of collodion. After drying, the collodion-mounted amino acids can easily be lifted off. The layer can also be scraped away directly; the system described by Schlicher [365] permits one to work almost without loss.* For the purpose of elution, 5 ml methanol containing 3% glacial acetic acid and 0.5% cadmium acetate are used [11, 13]. Collodion does not interfere, since it clearly dissolves in the elution agent. When the layer material has settled (possibly with centrifuging and filtering), photometry is performed at 500-510 mμ. With a layer thickness of 1 mm, for example, 2 μg serine furnishes an extinction of 0.040 [11]. Barrolier [13] found a linear relation between the quantity of amino acid and extinction value. The accuracy of the method is reported to be 2-5%.

The amino acids can be converted into the corresponding dinitrophenyl(DNP)-derivatives by means of 2,4-dinitrofluorobenzene (p. 126). The DNP-amino acids can be separated by thin-layer chromatography (p. 131), and a quantitative determination can be made by spectrophotometry without coloring. This method (measurement at 360 mμ; for DNP-Pro and DNP-Hypro at 385 mμ) was often used in paper chromatography [31]. The accuracy amounts to about 4-5% [39]. The quantitative evaluation of two-dimensional paper and thin-layer chromatograms is problematical *in situ* as well as after elution [205]. This is also indicated by the investigation of Filpowich [113], which states that the accuracy of a quantitative determination is 10-30% if *standard curves* are used for the evaluation.

Beale and Whitehead [19] have now developed an elegant ultramicro-method which permits a highly accurate ($\pm$ 3%) quantitative evaluation of two-dimensional chromatograms. Although this method was described for paper chromatography, its use is to be recommended particularly in thin-layer chromatography because of the very sharp separation of DNP-amino acids.

According to Beale and Whitehead [19], the amino acids are converted

---

* C. Roth Company, Karlsruhe, Germany.

with tritiated 2,4-dinitrofluorobenzene. The reaction mixture is treated with C^{14}-DNP-amino acids of known activity, and the DNP-amino acids are isolated by extraction (p. 128). The spots corresponding to the individual DNP-amino acids are scraped from the plate and the H^3 and C^{14} activity is measured (p. 35). The H^3 activity is a measure for the amino acid concentration; a decrease of the C^{14} activity shows the quantity lost during extraction and chromatography. The concentration of an amino acid is calculated by correcting the H^3 activity with the decrease of the C^{14} activity. Recently, it has become possible to determine the H^3 and C^{14} activities *simultaneously* and directly on the layer (p. 37) [436a]. Further references on quantitative analysis after elution are given in [478].

PART II

Thin-Layer Chromatography of Amino Acids, Peptides, and Related Compounds

Chapter 4

Thin-Layer Chromatography of Amino Acids

The practical advantage of thin-layer chromatography compared to paper essentially resides in the small amount of spreading of the sample spots (p. 2). This indicates that the differences are less absolute than gradual. The greater tendency for spreading of the spots probably must be attributed to the fiber structure of the filter paper. In thin-layer chromatography with the use of cellulose, aluminum oxide, or silica gel powder, this fiber structure no longer exists. Sudden spreading of the sample along the boundaries of fibers is no longer possible.

Thin-layers of inorganic or organic adsorbents serve for the chromatography of free amino acids. According to our present experiences, the success of separation depends primarily on the choice of solvent. Fundamental differences between inorganic and organic layers do not exist.

A. CHROMATOGRAPHY ON SILICA GEL LAYERS

1. Preparation of the Layer

Silica Gel Layers. The adsorbent together with distilled water is vigorously shaken for 30 sec. in a closed Erlenmeyer flask and is applied on 20 × 20 cm glass plates by means of an applicator (p. 4). For analytical chromatography, layer thicknesses of 0.25 mm are generally used (layer thickness during application). The ratio of adsorbent : water amounts to 1 : 2 with the use of Merck Silica Gel G, Merck Silica Gel H, and Macherey and Nagel Silica Gel G HR;* 2 : 3 with Woelm Silica Gel;† 2 : 5 with Camag Silica Gel D5 [302]. Recently we have used

---

* Macherey and Nagel & Co., Duren, Germany.
† M. Woelm AG, Eschwege, Germany.

Silica Gel G-1% Toronto starch* suspension to prepare the layer. The separation is not influenced by the presence of starch and the layers have a much greater strength.

The plates are dried in air overnight in horizontal position.

Buffered Silica Gel Layers. Silica gel (30 g) is mixed with 60 ml 0.2 M 1°-potassium phosphate/0.2 M 2°-potassium phosphate (1 : 1) to form a uniform suspension and are applied on 20 × 20 cm plates. The dry plates are activated at 110°C for 30 min. before use [267].

Supergel Layers. Supergel† (particle size < 0.071 mm) is washed twice with distilled water to neutral and is dried at 120°C for 24 hrs. The gel (9.5 g) and 0.5 gel plaster of Paris are mixed, screened, and shaken vigorously with double-distilled water (20 ml). The suspension is applied on a 20 × 20 cm plate and dried in air overnight.

2. Solvents and R_f-values

For the chromatography of amino acids on thin silica gel layers, many solvent mixtures can be used directly or with slight modifications from paper chromatography (Table 12).

For a one-dimensional development on Silica Gel G layers, solvents numbers 7 to 10 according to Brenner and Niederwieser [42] as well as numbers 12 and 13 according to Fahmy et al. [109] are especially appropriate. Solvent number 12, for example, can separate leucine and isoleucine even if these compounds are present together with many other amino acids. It is clearly evident from the R_f-values (Tables 13 to 15) that one-dimensional development does not permit a complete separation, particularly in the presence of complex mixtures of substances.

Mussini and Marcucci [265] chromatograph the n-butyl esters of amino acids on Silica Gel G layers with benzene/butanol (75 : 25) as solvent (Table 16). Only the esters migrate in this system; the free amino acids and the n-butyl-derivatives of the basic amino acids remain at the start. The latter can be developed, for example, with solvent number 14 (Table 16).

For the chromatography of the methyl- and n-amyl esters of amino acids, benzene/methanol (75 : 25) and benzene/pentanol (75 : 25) are suitable [266]. Since amino acid esters can be separated by gas chromatography, a rational combination of thin-layer and gas chromatography is indicated in this case.

As mentioned earlier, most of the solvents listed in Table 12 are already known from paper chromatography. Some of these solvent systems produce changes in the composition of the chromatographed amino

---

* Bender and Hobein, Zurich, Switzerland.

† Agfa/Woelfen.

Table 12. Solvents* for the chromatography of amino acids on silica gel layers (see text p. 65).

No.	Solvent		Author
1	Propanol/water	1 : 1	Nürnberg [279]
2	Phenol/water	5 : 2	
3	Ethanol/water	7 : 3	Mutschler and
4	Ethanol/ammonia	4 : 1	Rochelmeyer [267]
5	Ethanol/ammonia/water	7 : 1 : 2	
6	96% ethanol/water	7 : 3	Brenner and
7	Propanol/water	7 : 3	Niederwieser [42]
8	Butanol/glacial acetic acid/ water	4 : 1 : 1	
9	Phenol/water	75 : 25 w/w[1]	
10	Propanol/34% ammonia	7 : 3	
11	96% ethanol/34% ammonia .	7 : 3	
12	Methylethylketone/pyridine/ water/glacial acetic acid ..	70 : 15 : 15 : 2	Fahmy et al. [109]
13	Chloroform/methanol/ 17% ammonia	2 : 2 : 1	
14	Butanol/glacial acetic acid/ water	12 : 3 : 5	Checchi [62]
15	Phenol/water	75 : 25 g/g[2]	
16	Acetone/urea/water	60 : 0.5 : 40	Codern et al. [69]
17	Butanol/glacial acetic acid/ water	65 : 15 : 20	
18	Butanol/benzylalcohol saturated with water	1 : 1	
19	Butylalcohol/formic acid/ water	75 : 15 : 10	
20	Butanol/2-butanone/ 17% ammonia	5 : 3 : 1	
21	Collidine/water	25 : 44	
22	Pyridine/water	65 : 35	
23	Ethanol/ammonia	4 : 1	
24	Phenol/water	4 : 1 w/w	
25	Water/picoline	2 : 3	
26	Pyridine/isoamylalcohol/ water/diethylamine	10 : 10 : 7 : 0.3	
27	Ethylalcohol/ammonia/water	7 : 1 : 12	
28	Ethanol/butanol/ammonia/ water	4 : 1 : 2 : 3	
29	Lutidine/water	65 : 35	
30	Pyridine/isoamylalcohol/ water	35 : 35 : 30	
31	Butanol/2-butanone/water ..	2 : 2 : 1	

[1] 20 mg NaCN are added to 100 g.

[2] +0.1% benzoinoxime.

* Solvents in this and in all other tables are given in volume ratios unless stated as w/w.

Table 13. 100xR_f-values[1] of amino acids on buffered silica gel layers (see text p. 66). From Mutschler and Rochelmeyer [267].

Amino acid	100xR_f in solvent number		
	3	4	5
Lysine	05	05	10
Arginine	10	10	15
Histidine	25	65	75
Aspartic acid	40	15	40
Glycine	34	33	55
Serine	32	38	60
Proline	40	35	50
Alanine	50	50	65
Glutamic acid	50	25	55
Valine	65	70	82
Methionine	70	75	75
Leucine	75	75	85
Tyrosine	80	47	85

[1] Note that the R_f-values in this and the other tables depend on many factors (p. 39); consequently, they can be considered as only guide lines.

In this and all other R_f-tables of this book, the data are to be understood as applicable to the ascending technique and chamber saturation unless otherwise indicated.

Table 14. R_f-values of amino acids on Silica Gel G layers; compare p. 65.

Ref. No.	Amino acid	Abbrv.	100xR_f in solvent number							
			6[1]	7[1]	8[1]	9[1]	10[1]	11[1]	12[2]	13[3]
1	Alanine	Ala	47	37	22	29	39	40	51	50
2	β-alanine	β-Ala	33	26	22	30	30	29	35	36
3	α-aminobutyric acid	Abut	—	—	27	—	—	—	59	59
	α-aminoisobutyric acid	α-Aib	—	—	27	—	—	—	61	—
4	β-aminoisobutyric acid	β-Aib	—	—	25	—	—	—	48	47
5	γ-aminobutyric acid	γ-Abut	—	—	27	—	—	—	45	36
—	ε-aminocaproic acid	ε-Acap	—	—	34	—	—	—	68	—
—	α-aminocaprylic acid	Acy	66	65	59	69	58	60	143	—
6	Arginine	Arg	04	02	06	19	10	06	19	09
7	Asparagine	Asp(NH)$_2$	—	—	14	—	—	—	43	42
8	Aspartic acid	Asp	55	33	17	06	09	07	18	19
9	Citruline	Cit	—	—	—	—	—	—	—	47
10	Cysteic acid	CySO$_3$H	69	50	10	04	17	21	53	45
11	Cysteine	Cys	—	—	—	—	—	—	—	56
12	Cystine	(Cys)$_2$	39	32	09	12	27	22	18	56
13	Ethanolamine	Et-NH$_2$	—	—	—	—	—	—	—	39
14	Glutamic acid	Glu	63	35	24	10	14	15	32	34

Table 14—Continued

Ref. No.	Amino acid	Abbrv.	100xR_f in solvent number							
			6[1]	7[1]	8[1]	9[1]	10[1]	11[1]	12[2]	13[3]
15	Glutamine	Glu(NH$_2$)	—	—	15	—	—	—	—	49
16	Glycine	Gly	43	32	18	24	29	34	45	46
17	Histidine	His	33	20	05	32	38	42	18	55
18	Isoleucine	Ile	60	53	43	49	52	58	92	69
19	Leucine	Leu	61	55	44	48	53	58	100	70
20	Lysine	Lys	03	02	03	09	18	11	11	15
21	Methionine	Met	59	51	35	49	51	60	92	69
22	Methionine sulfone	MetO$_2$	—	—	—	—	—	—	66	63
23	3-methylhistidine ..	3-Mehis	—	—	—	—	—	—	—	59
—	1-methylhistidine ..	1-Mehis	—	—	—	—	—	—	09	—
—	Norleucine	Nleu	61	57	45	52	53	59	102	—
—	Norvaline	Nval	56	50	36	42	49	57	77	—
24	Ornithine	Orn	—	—	04	—	—	—	14	14
25	Hydroxyproline ..	Hypro	44	34	16	38	28	31	50	43
26	Proline	Pro	35	26	14	50	37	30	40	42
27	Phenylalanine	Phe	63	58	43	55	54	60	109	71
28	Phosphoethanoline	—	—	—	—	—	—	—	—	03
—	Sarcosine	Sar	31	22	12	37	34	31	32	—
29	Serine	Ser	48	35	18	20	27	31	47	41
30	Taurine	Tau	—	—	18	—	—	—	79	61
31	Threonine	Thr	50	37	20	26	37	40	51	56
32	Tryptophan	Try	65	62	47	63	55	58	122	69
33	Tyrosine	Tyr	65	57	41	47	42	51	107	62
34	Valine	Val	55	45	32	40	48	56	72	66

[1] Ascending technique; development distance: 10 cm [272].

[2] $R_{Leucine}$-values; horizontal technique; continuous flow chromatography (4 hrs.) [272].

[3] Ascending technique; development distance: 10 cm [346].

Table 15. R_f-values[1] of amino acids on Silica Gel G layers[2] (see text, p. 65). (Ascending technique; development distance: 15 cm). From: Codern et al. [69].

Amino acid (compare Table 14)	100xR_f in solvent number															
	16	17	18	19	20	21	22	23	24	25	26	27	28	29	30	31
Gly	72	16	22	22	07	10	74	33	21	90	23	55	62	61	26	10
Ala	75	20	34	31	09	18	76	50	24	88	34	68	67	69	31	13
β-Ala	61	22	18	27	06	20	67	34	28	79	17	49	50	53	28	12
Val	77	31	52	40	20	41	80	63	36	85	46	74	70	76	52	21
Abut	73	27	46	30	13	25	79	59	34	82	40	71	63	71	40	18
Nval	76	39	53	44	22	37	81	60	41	84	48	72	69	80	50	25
Leu	78	41	53	52	32	38	82	66	45	82	51	73	72	79	59	37
Ile	81	43	54	48	29	33	81	70	47	82	53	76	69	81	54	32
Nleu	80	46	53	58	34	31	83	61	52	84	57	80	71	82	60	41
Tyr	85	42	38	50	27	34	86	62	46	—	58	78	67	89	58	40
Ser	77	18	26	12	12	26	83	40	66	79	30	60	59	61	26	14

Table 15—Continued

Amino acid (compare Table 14)	16	17	18	19	20	21	22	23	24	25	26	27	28	29	30	31
Cys	—	08	—	05	07	12	—	40	—	—	—	75	72	—	—	07
Thr	81	23	56	17	24	10	87	48	20	84	34	74	68	68	39	13
Try	75	49	54	60	31	38	88	62	46	85	51	80	81	79	—	48
Pro	56	14	30	15	08	22	65	34	43	74	18	42	44	43	31	11
Asp	—	12	20	10	07	09	75	15	06	75	19	51	65	50	32	07
Asp(NH$_2$)	—	11	20	09	09	16	73	43	07	73	25	74	60	45	30	12
Glu	75	18	15	28	04	13	80	25	09	76	24	70	64	54	32	07
Arg	07	09	05	07	05	07	13	10	21	17	06	16	12	12	06	03
His	65	07	49	03	10	12	69	60	31	69	29	72	56	59	31	09
Lys	03	06	14	04	05	06	08	07	07	10	05	11	10	07	05	02

[1] The R$_f$-values of the Spanish authors partly contradict the usual experiences. This is true particularly for the leucines and for serine and threonine. The R$_f$-values, therefore, can be considered only as orienting data.

[2] Jost et al. [186] made use of phenol/water (9 : 3) for the separation of Tyr, N-methyl-Tyr, and O-methyl-Tyr; according to Peter et al. [319] Try and N-methyl-Try are separated in butanol/glacial acetic acid/water (3 : 1 : 1). Floss and Gröger [116] separate these compounds with butanol/glacial acetic acid/water (16 : 3 : 1). For the separation of His and methyl-His, compare Carisano [60] (p. 179).

Table 16. R$_f$-values of the butyl esters of amino acids on Silica Gel G layers (see text, p. 65).

Butyl esters of	100xR$_f$ value in	
	Benzene/n-butanol (75 : 25[1])	n-butanol/glacial acetic acid/water (12 : 3 : 5)[2]
Glycine	58	—
Alanine	54	—
Allo-hydroxyproline	31	—
α-aminobutyric acid	30	—
4-hydroxyproline	27.1	—
Norvaline	24.2	—
Valine	20	—
Norleucine	19.7	—
Leucine	18.6	—
Methionine	19.1	—
Isoleucine	18.8	—
Glutamic acid	15	—
Phenylalanine	14	—
Aspartic acid	13	—
Citrulline	08	—
Lysine	00	34
Histidine	00	36
Ornithine	00	35.5

[1] Development time: 1 hr.

[2] Development time: 3 hrs.

acids. Moses [260] and Opienska-Blauth [287] found that greater losses of amino acids occur in paper chromatography with phenolic solvents. This is particularly true for tryptophan; Opienska-Blauth [287] found that about 50% of 10 μg are destroyed (chromatography on Whatman No. 1 paper with phenol/water). Of the remaining amino acids, about 5% were decomposed [287]. Dent [90], Pikkarainen and Kulonen [321], and Edwards [105] report that methionine is partly oxidized in phenolic solvents. Dent [90] observed that about one half of the methionine is converted into methionine sulfoxide in phenol chromatograms. Cysteine and cystine are also partially oxidized as well as decarboxylated on paper [37] and on thin-layer chromatograms [109]. Losses on the one hand and side reactions on the other are partially prevented by reducing the development time, as is the case in thin-layer chromatography. Nevertheless, it is advisable to oxidize cysteine, cystine, and methionine with performic acid before chromatography (p. 115).

3. Separation of Amino Acids

As a rule, the separation of a larger number of amino acids is possible only by combining several solvents. Squibb [401] applies the samples on a preheated (65°C) silica gel plate and develops at 53°C with solvent number 8. After an intermediate drying, solvent number 9 is used to develop in the same direction under identical conditions. Mutschler and Rochelmeyer [267] use the combination of solvents numbers 3 and 4 for a two-dimensional development on buffered Silica Gel G layers; Nürnberg [279] uses Silica Gel G layers and solvents numbers 1 and 2.

Brenner and Niederwieser [42] investigated the reproducibility of a two-dimensional pattern of spots. They chromatographed two amino acid mixtures (A and B) separately and then together; when both two-dimensional chromatograms of A and B are superimposed, the same pattern is obtained approximately as if A and B were chromatographed together. In two-dimensional chromatograms, the conditions for intermediate drying must be standardized.

The combinations of solvent numbers 8 and 9 and of number 13 with 9 are particularly suitable for two-dimensional chromatograms [42, 45, 109].

Two-dimensional development with butanol/glacial acetic acid/water and with phenol/water. Brenner and Niederwieser proposed the combination of solvents numbers 8 and 9 [42]. Subsequently, these solvents were used in several laboratories for the separation of amino acids.

Baron [10] and Opienska-Blauth [289] reported on the separation of 17 and of 23 amino acids on Silica Gel G layers. Fig. 59a shows the spot pattern of Opienska-Blauth [289]. In our own experiments [302], a better separation could be observed in the basic amino acids (Figs. 59b and 1b).

Pataki [302] used different types of silica gel and found that the adsorbent had an influence on the separation effect of certain amino acid groups. The separation on starch-containing silica gel layers (cf. pp. 65-66) corresponds approximately to the data in Fig. 59b.

The separation of Lys, Arg, and His is most successful with Silica Gel G (Fig. 59b). These compounds move closer together on Silica Gel H layers as well as on MN-Silica Gel HR. All three *basic amino*

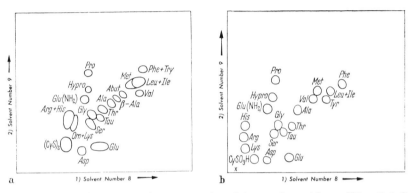

Fig. 59. Two-dimensional chromatography of free amino acids on Silica Gel G layers. The plate is dried (20 min. with airing) after the first dimension is developed.

a) Spot distribution according to Opienska-Blauth [289].

b) Spot distribution according to Pataki [302]. Amount of sample: 1 μg amino acid in a total of 1 μl 0.1 N HCl.

acids partially overlap on Woelm Silica Gel and on Camag Silica Gel D5. However, with a skillful use of a color reagent (p. 90), they can be distinguished. If ornithine, Lys, His, Arg, and Cit are present at the same time, solvents numbers 7 and 9 are combined [289].

The separation of Ser, Gly, and Tau is most successful with Silica Gel H or with Woelm Silica Gel [302]. The migration rate of Ser and Gly differs very little on Silica Gel G; Tau also is nearby. Ser, Gly, and Tau as well as Ser and Gly overlap on Camag Silica Gel D5 and MN-Silica Gel G-HR; however, a distinction is still possible [302].

Certain difficulties are presented by the identification of rapidly moving amino acids. Opienska-Blauth et al. [289] observed Phe + Try, Ile + Leu and Met + Val, or Ile + Leu + Met with the simultaneous separation of Val (Fig. 59a). We found [302] that the separation of Val, Met, Tyr, Leu(Ile), and Phe is excellent on Camag Silica Gel D5 and satisfactory on Silica Gel G. On MN-Silica Gel G-HR, Met + Val and Tyr + Leu appear in groups, while Met + Leu + Tyr appear together on Woelm Silica Gel. Nevertheless, their distinction is usually possible (p. 90 and Table 25).

In the chromatograms according to Fig. 59, Leu and Ile migrate together without exception. However, in a parallel experiment with the use of the continuous flow method (p. 19), they can be easily separated even if they are present together with all other protein amino acids + β-alanine + γ-aminobutyric acid (Fig. 60). In this method, Val and Tyr are also separated from the remaining amino acids (Table 13). Another method for the separation of leucine and isoleucine was published by Holbrook et al. [467].

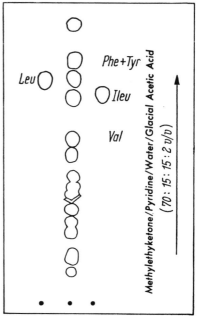

Fig. 60. Continuous flow chromatography for the separation of Leucine and Isoleucine in the BN-chamber (p. 16). Quantity of sample: 0.5 μg amino acid each in a total of 0.5 μl 0.1 N HCl; development time: 4½ hrs.; layer: Silica Gel G. For a definite identification of Leu and Ile, one sample each of these amino acids are chromatographed on the same plate. Performic acid oxidation (p. 115) is necessary if methionine is present. From Fahmy et al. [109].

The isomeric aminobutyric acids (α- and γ-aminobutyric acid and β-aminoisobutyric acid) are separated best on Silica Gel G layers with propanol/water (1 : 1) and phenol/water (7 : 3) [289].

Two-dimensional development with chloroform/methanol/water and phenol/water. Fahmy et al. [109] use Silica Gel G layers and solvents numbers 13 and 9 for the separation of 22 amino acids (Fig. 61a). Rokkones [346], Michl [251], and Neu [271] used the same chromatographic system; these authors investigated about 40 amino acids and related compounds. Fig. 61b shows the spot distribution of Rokkones [346] and Figs. 61c, d, and e show the spot charts of Michl [251] (see also Fig. 69b).

This method is very well suited for the detection of amino acids in protein hydrolysates (p. 116) as was also shown by Huber et al. [173], who substituted Silica Gel G by Supergel. Furthermore, it is also suited for the chromatography of amino acids in biological specimens (p. 173).

The two-dimensional spot patterns of Fahmy et al. [109] on the one hand and of Rokkones [346] on the other (Figs. 61a, b) are in good agreement with a few exceptions. Michl [251] observed a somewhat different spot distribution in some protein amino acids (Figs. 61c, d, e; compare also Fig. 69b).

In chromatograms such as Fig. 61, Leu and Ile migrate together; see Fig. 60 for their separation. Glu(NH$_2$), Cit, and β-Aib also are separated only incompletely (Fig. 61b). However, if a color reaction is used according to Moffat-Lytle (p. 92), then Glu(NH$_2$) can always be detected beside β-Aib. If β-Aib is present in very small concentrations, it is masked by Glu(NH$_2$). A polychromatic detection (p. 92) probably would also permit the identification of those amino acids which move together according to Michl (Figs. 61c, d, e); for example, Ala + Asp(NH$_2$) and Tyr + Phe.

The basic amino acids (Arg, Lys, Orn) are separated particularly well if solvent 13 is combined with phenol/15% aqu. formic acid (75 : 25 g/g) [74]. Rosetti used silica gel layers and different color reactions, which are carried out consecutively on the same plate, for the identification of amino acids [109].

Recently, silica gel-cellulose [459, 483] and silica gel carboxymethyl-cellulose [470] layers have also been used for the separation of amino acids. Eastman Kodak films have also been employed [481].

B. CHROMATOGRAPHY ON OTHER INORGANIC LAYERS

1. Aluminum Oxide and Celite Layers

Mottier [261] chromatographed the amino acids in the form of sodium salts on loose, non-adhering aluminum oxide layers. The best results are obtained with the use of Merck aluminum oxide (¾ hr. activation at 400°C, development time 8 hrs., continuous flow). Table 17 shows the R$_f$-values of 23 amino acids. Since loose layers are extremely fragile, this method has not become widely accepted.

Wohlleben [446] reported that adhering layers of Woelm basic aluminum oxide are suitable for the chromatography of amino acids. According to Rasteikiene and Prauskiene [333], the *aminodicarboxylic acids* are separated among each other as well as from the other amino acids on aluminum oxide layers with 2 N acetic acid as solvent.

Shasha and Whister [386] used Celite as adsorbent and isopropyl-alcohol/water (9 : 1) as a solvent. The development time is 25 min. for a distance of 12 cm (Table 17).

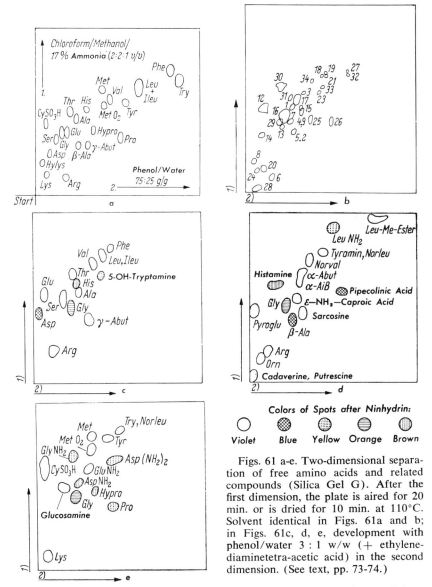

Colors of Spots after Ninhydrin:

Violet   Blue   Yellow   Orange   Brown

Figs. 61 a-e. Two-dimensional separation of free amino acids and related compounds (Silica Gel G). After the first dimension, the plate is aired for 20 min. or is dried for 10 min. at 110°C. Solvent identical in Figs. 61a and b; in Figs. 61c, d, e, development with phenol/water 3 : 1 w/w (+ ethylenediaminetetra-acetic acid) in the second dimension. (See text, pp. 73-74.)

a) Separation of protein amino acids + β-Ala + γ-Abut according to Fahmy et al. [109].

b) Relative position of amino acids and of other ninhydrine-positive substances according to Rokkones. The values refer to the reference number of the amino acids in Table 14. Note the good agreement between Figs. 61a and b. From Rokkones [346].

c, d, e) Spot charts according to Michl and Bachmayer [251]. Note that the position of some protein amino acids differs in *a and b*, respectively. From Michl and Bachmayer.

Table 17. R_f-values of amino acids on loose aluminum oxide layers [261] and on Celite layers [386].

Amino acid	A^1	B^2
Tryptophan	100	—
Proline	87	—
Phenylalanine	87	85
Norleucine	79	—
Isoleucine	75	—
Leucine	71	—
Norvaline	62	—
Tyrosine	56.5	—
Valine	55.5	77
Methionine	51	—
Alanine	47	51
Hydroxyproline	43.5	—
Histidine	39	—
Glycine	35	30
Threonine	34	—
Serine	32.5	32
Arginine	29	—
Glutamic acid	29	—
Cystine	29	—
Cysteine	20.5	15
Lysine	28	28
Aspartic acid	19.5	—

[1] $R_{tryptophan}$-values on Alox-layers: butanol/ethanol/water (3 : 2 : 2).

[2] R_f-values on Celite layers; solvent: isopropylalcohol/water (9 : 1).

2. Mixed Silica Gel/Kieselguhr Layers

Prey et al. [8] used 7.5 × 7.5 cm plates ("microplate method," p. 7) which were coated with silica gel/kieselguhr. The best layer combination proved to be silica gel/kieselguhr (4 : 11) suspended in water or silica gel/kieselguhr (1 : 1) suspended in 0.02 M sodium chloride. Table 18 shows the respective R_f-values. Silica gel/kieselguhr (1 : 1) layers and solvents of chloroform/methanol/ammonia (30 : 14 : 4), as well as butylacetate/glacial acetic acid/water (3 : 2 : 1), are suitable for two-dimensional chromatography (Table 18).

It is especially worthy of note that the time requirement for a two-dimensional chromatogram is about 30 min. It must be taken into consideration, however, that the load applied and its volume must be kept very small. The method is probably suited best for preliminary studies.

C. CHROMATOGRAPHY ON CELLULOSE LAYERS

1. Preparation of the Layer

MN-300-Cellulose Layers. Macherey and Nagel 300-Cellulose powder (15 g) are suspended in 90 ml water (von Arx and Neher [5] sus-

Table 18. R_f-values[1] of amino acids on silica gel/kieselguhr mixed layers. From: Prey et al. [8].

Amino acid	100xR_f in				
	A^2	B^3	C^4	D^5	E^6
Valine	20	28	48	43	28
Glutamic acid	00	03	05	31	02
Tyrosine	24	30	40	47	16
Isoleucine	30	34	60	47	33
Threonine	06	11	35	49	10
Phenylalanine	43	46	68	39	40
Serine	04	06	26	50	06
Histidine	06	10	18	10	10
Alanine	08	06	30	29	12
Arginine	00	00	06	12	00
Aspartic acid	00	00	04	12	00
Lysine	00	00	04	08	00
Leucine	36	36	60	51	34
Glycine	04	06	20	27	08
Tryptophan	46	42	61	58	32
Proline	06	05	29	29	14

[1] In the present chromatographic systems, a very distinct dependence of R_f on the development distance has been observed [8].

[2] Silica gel/kieselguhr (4 : 11); ethylacetate/ethanol/ammonia/water (30 : 10 : 2 : 3); development distance: 6 cm; time: 8 min.

[3] Layer, development distance, and time as in A; solvent: ethylacetate/ethanol/ammonia/water (15 : 7 : 2 : 1.5).

[4] Layer and development distance as in A; time: 9 min.; solvent: chloroform/methanol/ammonia (30 : 14 : 4).

[5] Silica gel/kieselguhr (1 : 1)/0.02 M NaCl suspension; butylacetate/glacial acetic acid/water (3 : 2 : 1); development distance: 6 cm; migration time: 12 min.

[6] Layer, development distance, and time as in D; chloroform/methanol/ammonia (30 : 14 : 4).

pend 8 g cellulose in 48 ml water plus 2 ml ethanol), are homogenized with an electric mixer, and applied on 20 × 20 cm plates (layer thickness during spreading: 0.25 mm). The layer is most advantageously dried overnight in air in a horizontal position.

Cellulose-TLC-layers. Cellulose powder-TLC* is shaken with the 10-fold quantity of water (e.g. 1 g + 10 g) in an Erlenmeyer flask for 10 min. and is applied on 20 × 20 cm plates (drying as above).

Cellulose-D layers. Cellulose powder D (Camag) (20 g) and 130 ml water are vigorously shaken for 60 sec. in an Erlenmeyer flask and applied on 20 × 20 cm plates with a thickness adjustment of 0.5 mm (drying as above).

MN-300-DEAE-Cellulose Layers. Macherey and Nagel 300-DEAE-cellulose powder (8 g) and 2 g MN-300-cellulose powder are homog-

* Serva Development Laboratory, Heidelberg, Germany.

Table 19. Solvents for the chromatography of amino acids on cellulose layers.[1]

No.	Solvent		Author
32	Butanol/glacial acetic acid/ water	4 : 1 : 5	Wollenweber [447]
33	Pyridine/methylethylketone/ water	15 : 70 : 15	
34	Methanol/water/pyridine ...	20 : 5 : 1	
35	Butanol/formic acid/water ..	15 : 3 : 2	
36	Propanol/8.8% ammonia ...	4 : 1	
37	Ethanol/butanol/water/ propionic acid	10 : 10 : 5 : 2	
38	Pyridine/isoamylalcohol/ water	7 : 6 : 6	Hörhammer et al. [170]
39	Isopropylalcohol/water	4 : 1	Dittmann [97]
40	Water-saturated phenol		Brandner and Virtanen [39]
41	Chloroform/methanol/ 17% ammonia	20 : 20 : 9	von Arx and Neher [5]
42	Butanol/acetone/diethylamine/ water	10 : 10 : 2 : 5	
43	Isopropylalcohol/99% formic acid/water	20 : 1 : 5	
44	sec.-butanol/methylethylke- tone/dicyclohexylamine/ water	10 : 10 : 2 : 5	
45	Phenol/water	75 : 25 w/w[2]	
46	Butanol/glacial acetic acid/ water	63 : 27 : 10	Myhill and Jackson [270]
47	Pyridine/water	4 : 1	Sjöholm [390]
48	Tert.-amylalcohol/methyl- ethylketone/water	3 : 1 : 1	Bujard [54]
49	Propanol/2 N ammonia	4 : 1	de la Llosa et al. [230]
50	Isopropylalcohol/2 N ammonia	4 : 1	
51	sec.-butanol/tert.-butanol/ methylethylketone/water .	1 : 1 : 1 : 1[3]	
52	Pyridine/glacial acetic acid/ water	30 : 10 : 7	
53	Methanol/pyridine/glacial acetic acid/water	80 : 4 : 1 : 20	

[1] Compare also solvents numbers 7, 12, and 13 in Table 12.

[2] Gas phase equilibrated with 3% aqu. NH_4OH.

[3] +0.5% diethylamine.

enized with 90 ml distilled water and applied on a glass support. Before application of the samples, it is most advantageous to develop with the specific solvent. With the use of acid-containing solvent mixtures, the DEAE-cellulose layer must be saturated with the respective acid (drying as above).

2. Solvents and R_f-Values

Many solvent mixtures have been proposed for the chromatography of amino acids on thin cellulose layers (Table 19).

According to Wollenweber [447], MN-300-cellulose layers and the Partridge mixture (solvent 32) are especially suited for a one-dimensional development. Table 20 contains the R_f-values of amino acids according to Wollenweber [447] and Bujard [54]. De la Llosa et al. [230] used MN-300-DEAE-cellulose layers and nine solvents. Table 21 contains the respective R_f-values.

3. Two-Dimensional Separation of Amino Acids

The separation of a complex mixture of amino acids on cellulose layers, just as on thin-layers of inorganic adsorbents, is successful only with the combined use of several solvents.

For the detection of most proteinogenic amino acids, the method of Hörhammer [170] or Sjöholm [390] can be used. According to Hörhammer [170], chromatography is carried out on cellulose-TLC layers with solvents numbers 32 and 38 (Fig. 62a). In this chromatographic system, Gly and Ser are not separated. The migration rate of Leu and Ile is also almost identical. Sjöholm [390] prefers MN-300-cellulose as the adsorbent and uses the solvent combination of numbers 32 and 47. Met, Val, and Phe, on the one hand, and Leu and Ile, on the other hand, do not separate (Fig. 62b).

Leu and Ile can be separated on MN-300-cellulose layers with solvent No. 12, if they are present alone [187]. In the presence of the other protein amino acids, solvents numbers 12 and 34 must be combined in order to resolve these two compounds [187]. With this solvent combination, it is possible to separate Ile, Phe, Tyr, Met, Val, and Pro among each other as well as from the other amino acids (Fig. 62c); in the groups Lys + Arg, Asp + Gly + Ser, as well as Thr + Glu + Ala, the components run with very similar migration rates.

An excellent separation of 28 amino acids, including Leu and Ile, is possible by the method of Bujard [54, 55]. Development in the first dimension is done in solvent number 13 and in the second dimension in number 34. MN-300-cellulose layers proved to be very useful as the stationary phase. In the chromatograms of Fig. 63a, Ala + Tyr and Leu + Ile do not separate. An additional two-dimensional chromatogram

Table 20. R_f-values[1] of amino acids on MN-300-cellulose layers (see text, p. 76).

Amino acid	100xR_f in solvent number		
	13[2]	32[3]	34[2]
Alanine	64	38	63
β-alanine	46	—	46
α-aminoadipic acid	29	—	61
γ-aminobutyric acid	53	—	48
Arginine	34	46	04
Asparigine	32	41	24
Aspartic acid	21	46	46
Citrulline	36	—	35
Cysteic acid	31	—	47
Cystine	24	14	22
Diaminopimelic acid	19	—	13
Glutamic acid	26	35	62
Glutamine	39	—	34
Glycine	42	29	46
Histidine	48	—	27
Homoarginine	42	—	05
Homocystine	34	—	20
allo-hydroxylysine	51	—	03
Isoleucine	90	67	81
Leucine	85	70	82
Lysine	56	42	05
Methionine	81	50	66
Methionine sulfone	67	—	49
Methionine sulfoxide	54	—	45
3-methylhistidine	63	—	44
1-methylhistidine	63	—	34
Ornithine	44	—	03
Oxyproline	46	—	46
Proline	75	—	64
Phenylalanine	89	—	71
Serine	49	46	46
Taurine	71	—	52
Threonine	51	—	53
Tryptophan	68	60	50
Tyrosine	62	46	60
Valine	85	54	72

[1] Myhill and Jackson [270] separate Pro, Hypro and their nitroso-derivatives in solvent number 46. Pro (70); Hypro (0.36) N-nitrose-Pro (64); N-nitroso-Hypro (35) (R_f-values in parentheses); for the detection of creatinine, see pp. 177-179.

[2] According to Bujard [53]; compare Table 12.

[3] According to Wollenweber [447].

(solvents numbers 48 and 43) separates these groups and other amino acids which otherwise migrate rapidly (Fig. 63b).

If Hylys is present, this compound migrates between Orn and Lys

Table 21. R_f-values of amino acids on MN-300-DEAE-cellulose layers (see text, pp. 77, 79). From: de la Llosa et al. [230].

Amino acid	100xR_f in solvent number								
	12[1]	32[1]	34	47[2]	49[3]	50[3]	51[3]	52[2]	53[1]
Leucine	36	55	75	54	52	46	42	65	70
Isoleucine	31	51	73	53	48	40	39	63	70
Phenylalanine	42	45	62	52	44	35	40	61	62
Valine	21	39	69	45	40	33	28	56	66
Methionine	30	37	62	48	35	29	30	52	59
Proline	14	26	66	33	30	27	22	36	58
Tryptophan	—	36	—	46	28	—	35	—	45
Tyrosine	31	29	61	48	22	—	28	57	54
Alanine	10	23	55	30	22	20	17	36	57
Threonine	11	19	55	33	14	17	22	38	50
Serine	05	14	50	28	11	08	11	26	45
Glutamine	06	15	47	21	08	20	11	26	43
Glutamic acid	00	14	03	00	02	03	02	26	14
Glycine	07	17	47	18	12	11	10	22	42
Histidine	04	22	41	21	10	11	11	22	48
Arginine	04	24	71	27	14	10	11	23	62
Lysine	02	20	69	16	11	11	06	18	60
Asparagine	05	13	36	16	07	07	08	18	36
Aspartic acid	00	09	02	00	02	03	04	—	07
Cystine	—	06	—	—	01	—	—	—	—

[1] Before spotting of samples: flushing run with the solvent or flushing run with methanol containing 20% glacial acetic acid; compare Table 12.

[2] Before spotting of samples: flushing run with the solvent.

[3] Before development, equilibration for 1 to 2 hrs. with the solvent or moving the plate over an open bottle of ammonia.

and may overlap with the former amino acid. Glu(NH₂) and Cit are separated only incompletely; furthermore, Tau is near Try.

Dittman [97, 98] chromatographed 48 amino acids and related compounds on MN-300-cellulose layers. He developed first with solvent number 7 (12 cm) and then with number 39 in the second dimension (10 cm). Finally, solvent number 7 was again allowed to ascend in the first dimension (all runs without chamber saturation). Table 22 contains the R_f-values obtained from two-dimensional chromatograms [97, 98]. Cf. references [466, 472] dealing with other methods for the separation of amino acids on cellulose layers.

4. Multidimensional Separation of Amino Acids

Von Arx and Neher [5] developed a combination of four solvent mixtures and different color reactions for the separation and characterization of 52 amino acids ("multidimensional technique," p. 22).

The samples are spotted on 6 MN-300-cellulose plates at a distance

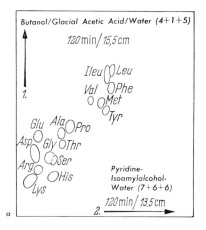

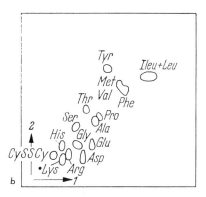

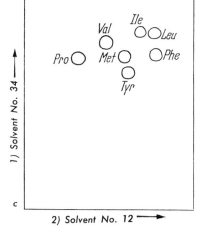

Fig. 62. Two-dimensional chromatography of the free amino acids on cellulose layers.

a) Spot pattern on cellulose-TLC layers (see p. 77). After the first dimension, drying for 10 min. with a fan. From Hörhammer et al. [170].

b) Spot distribution on MN-300-cellulose layers (see p. 77). First dimension: butanol/glacial acetic acid/water (4 : 1 : 5); second dimension: pyridine/water (4 : 1). After the first run, drying of the layer for 30 min. at 70°C. CySSCy = (Cys)₂. From Sjöholm [390].

c) Separation of "fast-moving" amino acids including Leu and Ile. Prepared according to Jutisz and de la Llosa [187]. If other amino acids are present, they do not interfere with separation; their resolution is incomplete, however. (Try was not investigated.) (Compare Table 12.)

of 2.5 cm from the edge and development subsequently proceeds with solvent number 42. The "multiplate tanks" of the Shandon Company (p. 12) serve for chromatography. Since the process must be carried out without saturation, the plates and solvent are introduced into the chamber at the same time (distance between start and immersion line: 1.5-2 cm). After drying (5-10 min. at 90°C), a mixture of 0.5 μg Gly,

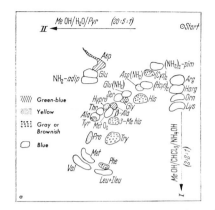
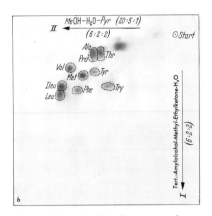

Fig. 63. Two-dimensional separation of amino acids and related compounds on MN-300-cellulose layers.

a) First dimension: Methanol/chloroform/17% ammonia (2 : 2 : 1); second dimension: methanol/water/pyridine (20 : 5 : 1). Note the excellent separation of basic amino acids (Bujard [54]). $(NH_2)_2$-pim = diaminopimelic acid; NH_2-adip = α-aminoadipic acid; otherwise compare Table 14.

b) Separation of the "fast-moving" amino acids including Leu and Ile. Note the good separation of Val, Met, Tyr from Ph and Try, and the separation of Leu and Ile (Bujard [55]). First dimension: tert.-amylalcohol/methylethylketone/H_2O (6 : 2 : 2); second dimension: methanol/H_2O/pyridine (20 : 5 : 1).

1 μg Tyr, and 1 μg Nleu in 1 μl water is dropped on the points indicated by "test" in Fig. 64. Two plates each are developed with solvent number 43 in two chambers, one with solvent number 44 and one with solvent number 45 in the second dimension to the upper edge of the plate (conditions as in the first dimension). The plates are now dried for 20 min. at 90°C and are sprayed with different color reagents. One plate each of the three combinations is sprayed with Ninhydrin/collidine reagent (p. 92); one chromatogram of the combination of solvents 42-43 is treated with isatin (p. 93); finally, the others are treated with specific reagents such as Paulys reagent (p. 93) or nitroprusside-ferricyanide reagent or Reindel-Hoppe reagent.

In Fig. 64, the three two-dimensional chromatograms are shown.

In Scheme I (see page 211), the amino acids are arranged in the order of increasing R_f-values in solvent number 42. For each compound, the R_f-values in the solvents used for the second dimension (numbers 43, 44, and 45) are listed.

Any shifts of R_f-values in the second dimension can be controlled and recognized by simultaneous chromatography of test substances (Fig. 64). The assignment of amino acids also is facilitated by different color reactions (see above).

However, this excellent method should be used only with very complex amino acid mixtures. Other methods are available for simpler

Table 22. R_f-values[1,2] of amino acids in two-dimensional chromatograms on MN-300-cellulose layers (see pp. 77, 79). From Dittmann [97, 98].

Amino acid	R_{f1}[3]	R_{f2}[4]
Ornithine	08 ± 01	00
Histidine	10 ± 02	01 ± 01
Histamine	10 ± 02	01 ± 01
Lysine	10 ± 02	01 ± 01
1-methylhistidine	10 ± 02	02 ± 01
Arginine	19 ± 03	01 ± 01
Cystine	22 ± 01	02 ± 01
Asparagine	34 ± 02	08 ± 01
Glutamine	41 ± 03	12 ± 01
Aspartic acid[5]	45 ± 02	11 ± 02
Citrulline[5]	48 ± 03	12 ± 01
Glycine	49 ± 04	17 ± 01
Taurine	50 ± 03	26 ± 01
Serine[5]	52 ± 03	17 ± 01
Glutamic acid	55 ± 02	17 ± 01
Hydroxyproline	62 ± 03	23 ± 01
Alanine	63 ± 02	27 ± 01
Threonine	64 ± 03	31 ± 01
Sarcosine	65 ± 04	25 ± 02
Proline	66 ± 02	32 ± 01
Ethanolamine	66 ± 02	38 ± 04
Glucosamine	67 ± 02	24 ± 02
β-alanine	68 ± 04	30 ± 01
δ-aminolevulinic acid	71 ± 01	30 ± 02
Tyrosine	71 ± 02	34 ± 03
β-aminoisobutyric acid	74 ± 03	36 ± 00
α-aminobutyric acid	75 ± 02	38 ± 02
β-amino-n-butyric acid	75 ± 02	38 ± 02
γ-aminobutyric acid	75 ± 02	38 ± 02
Valine	78 ± 02	45 ± 03
α-aminoisobutyric acid	79 ± 03	43 ± 01
Tryptophan	80 ± 02	34 ± 01
Methionine	83 ± 02	47 ± 01
Phenylalanine	84 ± 02	51 ± 02
Norvaline	85 ± 02	54 ± 02
Leucine	88 ± 01	64 ± 03
Isoleucine	89 ± 03	62 ± 04
3,5-diiodotyrosine	91 ± 01	51 ± 02

[1] Averages from 4 determinations ± deviation.

[2] The chromatogram was developed with solvent number 7 and after intermediate drying (30 min. in air) with number 39 (second dimension). After another drying period, chromatography was repeated in the first dimension with number 7 (development distances 12 cm; without chamber saturation; compare Tables 12 and 19).

[3] R_f in the first dimension.

[4] R_f in the second dimension.

[5] Form secondary spots.

problems, for example, detection of amino acids in protein hydrolysates (p. 72, 73, and 83).

D. ELECTROCHROMATOGRAPHY

Numerous devices have been described for the electrophoretic separation of mixtures of substances on thin adsorbent layers (p. 33). Honegger [163], Pastuska [314], Kaplan and Schneider [188], Nybom [281], Hannig [151], Stegemann [413, 414], and Bieleski and Turner [459] dealt with the separation of amino acids.

In 1953, Biserte et al. [32] showed that amino acid mixtures could be pre-separated by paper electrophoresis (at pH = 3.9; 400 V; 5 hrs.). The pre-separation already permits a characterization of CySO₃H, Asp,

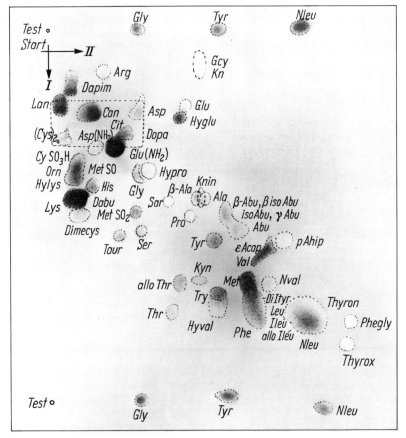

Fig. 64a. First dimension: solvent number 42 (designated by 1 in the figure); second dimension (after drying for 5-10 min. at 90°C): solvent number 43 (II). See p. 87.

Glu, β-Aib, β-Ala, γ-Abut, and Aad [33, 338, 339]. The basic and neutral amino acids each form one group; their separation takes place by electrophoresis or chromatography [32, 33, 338]. Subsequently, Montant and Tourze-Poulet [257] attempted to apply this technique to cellulose layers of 3-4 mm thickness (250 V; 8 mA; 24 hrs.), and they achieved a similar pre-separation as Biserte et al. The method is suitable for a preparative amino acid separation; however, it did not become widely accepted probably because of the long development time and the difficulties in layer preparation.

In 1962, Honegger [163] reported on a successful electrochromatographic separation of a few amino acids and amines. The Swiss author used Silica Gel G/citrate layers and worked according to the data of Table 23 (No. I; first dimension, electrophoresis) (compare also Stegemann [413, 414], p. 181 and No. IV in Table 23).

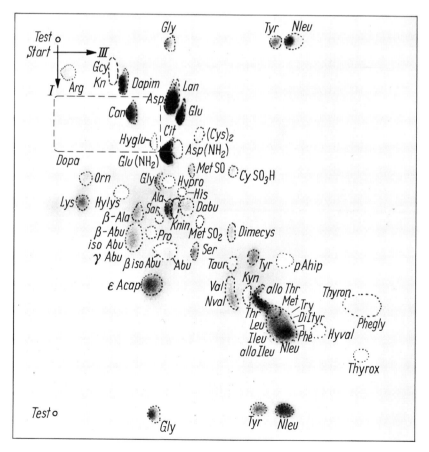

Fig. 64b. First dimension: solvent number 42 (I); second dimension (after drying): number 44 (III). See p. 87.

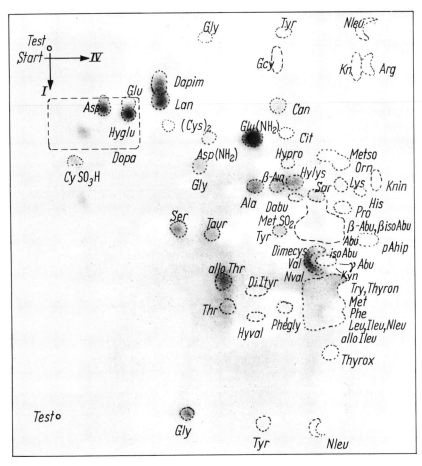

Fig. 64c. First dimension: solvent number 42 (I); second dimension (after drying): solvent number 45 (IV). From von Arx and Neher [5].

Fig. 64. Multidimensional separation of the free amino acids and numerous related compounds according to von Arx and Neher (compare scheme I). The abbreviations can be found in Tables 14, 25, and Scheme I (MN-300-cellulose layers; compare text). All runs *without* chamber saturation: (See text pp. 38, 41, 81).

Pastuska [314] also used Silica Gel G layers (suspension in 3% borax solution) and conducted the separation at pH = 2 (2 hrs., 10 V/cm). Under these conditions, all the amino acids investigated migrated to the cathode. Kaplan and Schneider [188] performed the electrophoresis on MN-300-cellulose layers at pH = 1.9 (2 hrs.; 500 V; 10 mA) and separated alanine and asparagine.

Nybom [281] combined electrophoresis (first dimension) with chromatography (second dimension) and obtained a satisfactory separation for a number of amino acids. Table 24 contains the data for the electrophoretic and chromatographic migration rates of 21 amino acids.

Table 23. Electrochromatography of amino acids.

No.	Electrophoresis		Chroma-tography	Author	Remarks
	Buffer	Conditions			
I	2 N acetic acid/0.6 N formic acid 1 : 1 (pH = 2)	460 V; 12.6 mA, 160% humidity; 1 hr.	Butanol/glacial acetic acid/ water 4 : 1 : 1	Honeg-ger [163]	Silica gel prepared with 0.1 M citrate buffer (pH=3.8) (0.25 mm)
II	0.7% (v/v) formic acid	400 V; 15 min.	Butanol/formic acid/water 4 : 1 : 2 Propanol/pyri-dine/water 5 : 1 : 2 Methylethyl-ketone/formic acid/water 24 : 1 : 6	Nybom [281]	MN-300-cellulose (0.35 mm)
III	25 ml formic acid + 78 ml acetic acid + water to 1 l	400 V; 1 hr.; 90-200% hum.	Chloroform/ methanol/17% ammonia 2 : 2 : 1	Hannig [151]	Cellulose or silica gel layers
IV	240 ml gla-cial acetic acid + 60 ml pyridine + water to 5 l (pH = 3.9)	1200 V/20 cm; 25 min.	Butanol/glacial acetic acid/ water 4 : 1 : 1	Stege-mann and Lerch [414]	30 g Silica H or 15 g MN-300-cellulose + 80 ml buffer[1]

[1] Pyridine/glacial acetic acid/water 20 : 9.5 : 970.

The method could be used for the detection of amino acids in protein hydrolysates (Fig. 70) and in biological material (Figs. 83 and 84).

The separation of 14 amino acids has been reported by Hannig [151]. Ionophoresis is carried out in the first and chromatography in the second direction. The experimental conditions are given in Table 23 (No. III). Either silica gel or cellulose can serve as the support.

Fig. 65 shows the electrochromatographic separation of amino acids according to Bielski and Turner [459].

E. DETECTION OF AMINO ACIDS

The Ninhydrin reaction is generally used for the detection of amino acids in thin-layer chromatography. A definite identification admittedly

Table 24. Ionophoretic migration rate and R_f-values of amino acids on MN-300-cellulose layers (see text p. 40). From Nybom [281].

Amino acid	$U_{proline}$[1]	R_{fA}[2]	R_{fB}[3]	R_{fC}[4]
Tryptophan	84	62	48	56
Hydroxyproline	86	40	24	14
Aspartic acid	93	37	18	19
Tyrosine	94	57	46	51
Proline	100	51	33	26
Phenylalanine	100	66	58	55
Glutamine	103	34	14	12
Asparagine	105	27	11	09
Glutamic acid	105	43	23	25
Methionine	107	60	51	50
Threonine	113	42	26	22
Serine	115	35	18	13
Leucine	115	70	64	57
Valine	120	60	48	46
Alanine	132	47	28	27
Glycine	145	37	16	16
Arginine	197	33	00	07
Histidine	203	27	06	04
Lysine	209	29	00	04
γ-aminobutyric acid	216	54	22	37
β-alanine	222	48	20	30

[1] Ionophoretic migration rate referred to proline; compare Table 23, No. II.

[2] Butanol/glacial acetic acid/water 4 : 1 : 2; MN-300-cellulose.

[3] Propanol/pyridine/water 5 : 1 : 2; MN-300-cellulose.

[4] Methylethylketone/formic acid/water 24 : 1 : 6; MN-300 cellulose.

Fig. 65. Map of amino acids, sugars, and organic acids, separated by TLE (buffer = 17 ml 90% HCOOH + 57 ml CH₃COOH + 926 ml water, pH 2.2; 20 min, 1000 V, 30 mA) and then chromatography in methylethylketone/pyridine/water/acetic acid (75 : 15 : 15 : 2, v/v) for 100 min followed by n-propanol/water/n-propylacetate/acetic acid /pyridine (120 : 60 : 20 : 4 : 1, v/v) for 5 hr, on Silica Gel H–MN-300 cellulose (1 : 2.5, w/w). Abbreviations: Met 1 = methionine sulfoxide, Met 2 = methionine sulfone. Pyrrolidonecarboxylic acid lies in approximately the same position as citric acid. For remaining abbreviations, cf. ref. [459].

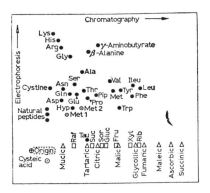

is not possible because many other compounds also react with Ninhydrin. These "erroneously Ninhydrin-positive" compounds, including some non-nitrogenous substances [317, 364, 451, 452], can lead to misinterpretations. The accuracy of a chromatographic characterization is highly increased if a few specific color reactions are carried out (for the specific detection of hydroxyproline, see Stegemann [413a]; the sensitivity amounts to 10 ng after 24 hrs.). Amino acids resolved on chromatoplates can also be identified by reflectance spectrophotometry used in conjunction with R_f-values and visual observation; e.g., Ninhydrin color [463].

1. The Ninhydrin Reaction

The reaction of amino acids and peptides with triketohydrindene hydrate (Ninhydrin) has been known for more than fifty years. The mechanism of this reaction has not yet been fully determined; background information concerning the subject can be found in Caldin [59].

a) *Silica Gel Layers.* The detection of amino acids on silica gel layers is carried out as follows according to Brenner et al. [46]. The carefully dried layer is sprayed with the Ninhydrin solution (0.3 g Ninhydrin + 100 ml butanol + 3 ml glacial acetic acid). The plate is now held over a hot-plate until the color development just begins [42]. The gradual deepening of the color spots can be observed in transmitted light. Some amino acids thus appear almost as dots and are immediately marked with a sharp pencil. In this manner it is often possible to recognize individual components in spots which become confluent subsequently [42]. The amino acids also differ in the rate with which they form color. Subsequently, the plate is heated to 60°C for 30 min. or to 110°C for 10 min.

According to Opienska-Blauth [289], the spots formed at room temperature are marked, and the plate is heated to 60°C. The identification of many amino acids is already possible in the cold; some less sensitive compounds appear only after heating.

The method of Barrolier [11, 13, 15] is particularly suited for the quantitative determination of amino acids (p. 60): The plate is in *horizontal* position and is sprayed with the reagent solution (100 mg cadmium acetate + 10 ml water + 5 ml glacial acetic acid + 100 ml acetone + 1 g Ninhydrin). The layer must be completely impregnated but should not "float." Approximately 50 ml of spray solution are required for a 20 × 20 cm plate. Color development takes place at room temperature (about 24 hrs.) in the absence of light.

Table 25 shows the data concerning the detection sensitivity of amino acids.

b) *Cellulose Layers.* According to von Arx and Neher [5], the layer is dried at 90°C and sprayed with the reagent solution (1 g Ninhydrin + 700 ml abs. ethanol + 29 ml 2,4,6-collidine + 210 ml glacial acetic acid). Drying now takes place at 90°C until the spots become visible;

Table 25. Detection of amino acids on thin-layer chromatograms (detection sensitivity and colors).

Amino acid	Ninhydrin reaction		Chlo-rine-tolidine react.[3]	Isatin react.[4]	Moffat-Lytle react.[5]
	Silica gel[1]	Cellu-lose[2]			
Alanine	0.05	0.05	—	violet	pink
β-alanine	0.06	0.05	+	lilac	violet
α-aminobutyric acid ...	0.04	0.05	(+)	lilac	pinkish-red
α-aminoisobutyric acid .	—	0.05	(+)	yellow	—
β-aminobutyric acid ...	—	0.05	+	yellow	—
β-aminoisobutyric acid .	—	0.05	+	lilac	dark yellow
γ-aminobutyric acid ...	—	0.2	+	lilac	yellowish-pink
ε-aminocaproic acid	—	1.0	+	pink	—
Arginine	0.06	0.2	+	violet	pink
Asparagine	—	2.0	+	pink	brownish-orange
Aspartic acid	0.2	0.05	(+)	violet	gray
Citrulline	—	0.1	+	pink	pink
Cysteic acid	0.1	0.05	(+)	yellow	reddish-gray
Cysteine	—	—			reddish-gray
Cystine	—	1.0	(yellow)	pink	reddish gray
α, γ-diaminobutyric acid	—	2.0	+	pink	—
α, α-diaminopimelic acid	—	2.0	(gray)	red	—
Dihydroxyphenylalanine	—	10.0	(yellow)	lilac	—
Diiodotyrosine	—	5.0	+	lilac	—
Glutamine	—	0.5	(+)	pink	yellowish-pink
Glutamic acid	0.4	0.05	—	lilac	pink
Glycine	0.006	0.05	+	pink	orange
Histidine	0.5	0.3	(gray)	lilac	purple
Hydroxyglutamic acid ..	—	0.05	(+)	pink	—
Hydroxylysine	—	0.5	+	lilac	—
Hydroxyproline	0.1	1.0	(+)	blue	yellowish-gray
β-hydroxyvaline	—	0.5	(+)	pink	—
Isoleucine	0.2	0.1	(+)	pink	pink
allo-isoleucine	—	0.1	(+)	pink	—
Creatine	—	—	(+)	yellow	—
Creatinine	—	—	+	yellow	—
Kynurenine	—	0.5	(yellow)	pink	—
Lanthionine	—	1.0	(pink)	orange	—
Leucine	0.2	0.5	—	pink	pink
Lysine	0.03	0.5	+	red	yellowish-red
Methionine	0.4	0.5	—	pink	light orange
Methioninesulfone	0.2	0.2	(brown)	pink	gray
Methioninesulfoxide ...	—	0.2	(brown)	pink	—
Norleucine	—	0.5	—	lilac	—
Norvaline	—	0.05	(+)	red	—

Table 25—Continued

| Amino acid | Ninhydrin reaction | | Chlo-rine-tolidine react.[3] | Isatin react.[4] | Moffat-Lytle react.[5] |
	Silica gel[1]	Cellu-lose[2]			
Ornithine	—	0.2	+	red	dark pink
Phenylalanine	0.2	0.2	(yellow)	lilac	orange-brown
α-Phenylglycine	—	5.0	(brown)	yellow	—
Proline	0.5	0.5	(+)	blue	yellow
Sarcosine	—	0.1	+	yellow	—
Serine	0.1	0.2	yellow	orange	yellowish-red
Taurine	—	0.2	+	yellow	gray
Threonine	0.1	0.5	(+)	pink	yellowish-red
allo-threonine	—	0.5	(+)	pink	—
Thyronine	—	2.0	—	brown	—
Thyroxine	—	2.0	(+)	yellow	—
Tryptophan	0.5	0.5	(brown)	lilac	blue
Tyrosine	0.1	0.5	(pink)	red	yellowish-red
Valine	0.2	0.05	(+)	pink	pink

[1] According to Fahmy et al. [109] (compare Fig. 61a and p. 90).

[2] According to von Arx and Neher [5] (compare Fig. 98 and p. 92). Creatinine and creatine do not react with Ninhydrin.

[3] According to von Arx and Neher [5]: + = bluish-black color; (+) = weak reaction. Data for 0.5-5 μg amino acid each (see text, p. 107).

[4] According to von Arx and Neher [5]; data for 0.5-5 μg amino acid each (see below).

[5] Colors of the amino acids on chromatograms according to Fig. 61: compiled according to Rokkones [346] and author's own findings [306].

these must be immediately marked in order to facilitate the analysis. After a few hours at room temperature, the colors become more intense and may coalesce. The limits of detection of amino acids are shown in Table 25.

c) *Polychromatic Detection.* The Moffat-Lytle reagent [254] is suitable for a polychromatic detection of amino acids. The plates are heated to 110°C for 10 min., cooled, and sprayed with the color reagent.

Solution I: 50 ml 0.2% Ninhydrin solution in abs. ethanol + 10 ml glacial acetic acid + 2 ml 2,4,6-collidine.

Solution II: 1% Cu(NO₃)₂ · 3H₂O solution in abs. ethanol.

Solutions I and II are mixed in a ratio of 50 : 3 just before use. For the treatment after spraying, see a), p. 90.

Many amino acids exhibited characteristic colors which are listed in Table 25.

2. Non-Destructive Detection

As mentioned above, the Ninhydrin reaction is not a specific method for the detection of amino acids. Consequently, in the investigation of complex mixtures, it is often desirable to verify the identity of the spots after elution by a rechromatography. If necessary, the structure of the eluted substances can be verified by physical methods (e.g. spectra).

According to Pataki [300, 300a], 2,4-dinitrofluorobenzene (DNFB) is suitable for the non-destructive detection of amino acids. In this color reaction, yellow dinitrophenyl (DNP) amino acids are formed (p. 126) and are visible in daylight as well as in the UV region. It has not yet been determined which derivatives of histidine, tyrosine or lysine, etc. are formed. The DNP-amino acids can be eluted; they are then rechromatographed in a suitable solvent (pp. 130-131). In order to record spectra, larger quantities of substance are isolated with the aid of preparative thin-layer chromatography (pp. 28 ff).

In this connection, reference should be made to the study of Cherbuliez [17], according to whom IR-spectra can be recorded of μmole-quantities.

The dried plate is sprayed with a buffer solution (8.4 g $NaHCO_3$ + water + 1 N NaOH pH = 8.8 + water to 100 ml) and with a 1% (g/v) methanolic DNFB solution. The layer is wiped off over a width of 5 mm on both edges of the plate. Two polyethylene strips of suitable width are placed on the wiped edges, and the layer is covered with a second glass plate and heated for 1 hr. to 40°C in darkness. The carrier plate is cooled and exposed to HCl vapors. After 10 min., it is briefly dried and the spots are marked. Approximately $10^{-2} - 10^{-3}$ μmole amino acid can be detected.

3. Other Color Reactions

a) *Chlorine/Tolidine Reaction.* The NH-CO groups are especially sensitive to the chlorine/tolidine reaction. The procedure is described on pp. 107, 108. Table 25 contains some data on the coloring of amino acids with this method; it must be noted, however, that the required concentrations are relatively high.

b) *Isatin Reaction.* In the isatin reaction [14], the colors developed by some amino acids are more differentiated than in the Ninhydrin reaction.

Spray solution I: 1 g isatin + 1.5 g zinc acetate + 1 ml glacial acetic acid + 95 ml isopropylalcohol + 5 ml water.

Spray solution II: 1 g isatin + 1.5 g zinc acetate + 1 ml pyridine + 100 ml isopropylalcohol.

The plate is sprayed with either solution I or solution II. According to von Arx and Neher [5], the heavily sprayed layer is dried for 20 hrs. at 20°C. Table 25 gives information on the colors developed by individual amino acids [5].

c) *Paulys Reagent.* Histidine, tyrosine, thyroxine, and dihydroxyphenylalanine form yellow to reddish colors. Diiodotyrosine and creatine have only weak reactions [5].

Solution I: 0.4 M sodium sulfanilate in water.

Solution II: 0.4 M sodium nitrite in water.

Before use, 1 vol. solution I + 1 vol. solution II + 8 vols. 0.25 N HCl + 10 vols. 2 N soda solution are mixed.

d) *Folin Reagent.* Sodium β-naphthoquinone-4-sulfonate (0.2 g) + 100 ml 5% sodium carbonate solution are mixed and allowed to stand for 10 min. The dried plate is now uniformly sprayed with the reagent. It should not be sprayed too heavily or the cellulose will turn slightly yellow. A rapid identification of individual amino acids is achieved by the characteristic color (Müting [268]). The amino acids form light blue, grayish-green to pink-red spots on a white background; the limit of detection is about 0.5-1 μg [170].

e) *Sodium Nitroprusside/Potassium Ferricyanide Reagent.* One volume each of 10% sodium nitroprusside and 10% potassium ferricyanide solution are mixed with 3 vols. dist. water. The solution is allowed to stand for 30 min. and the layer is then sprayed [5]. A red color is formed by, for example, arginine, creatine, and creatinine.

f) *Sakaguchi Reaction.* This reaction is specific for the guanido group. Arginine and homoarginine turn pink-red [54].

The chromatogram is sprayed with a mixture of a 16% urea solution and an 0.2% ethanolic α-napthol solution (5 : 1). The layer is dried at about 40°C and subsequently sprayed with a solution of 3.3 ml bromine in 500 ml 5% sodium hydroxide solution.

On the use of different color reagents successively on the same plate, cf. Rosetti [479].

F. THIN-LAYER CHROMATOGRAPHY OF IODOAMINO ACIDS

1. Solvents and Separation Effects

Layers of silica gel [247, 366, 412], cellulose [162, 315], and Dowex [26] are suitable for the chromatography of iodoamino acids (Table 26). Schneider and Schneider [366] use Silica Gel G as the stationary phase for the separation of monoiodotyrosine (MIT), diiodotyrosine (DIT), diiodothyronine (T-2), triiodothyronine (T-3), and thyroxine (T-4). The best separation is obtained with solvent number 54 (Table 26).

The iodoamino acids migrate in the sequence of MIT, DIT, T-2, T-3, and T-4 (development distance 11.5 cm; 1 hr. development); MIT has an R_f-value of about 0.25; T-4 has an R_f-value of about 0.46. In this chromatographic system, nitrogen-free aromatic iodo-acids migrate with the solvent front; for example, diiodohydroxyphenylpyruvic acid, di-iodo-, triiodo-, and tetraiodothyroacetic acid. Iodine-free amino acids generally have lower R_f-values than MIT. Iodine, which does not produce a sharp spot, migrates between MIT and DIT. T-2, T-3, and T-4 can be separated from each other as well as from the iodotyrosines by means of solvent number 55 ($R_{fT-2} \sim 0.3$; $R_{fT-3} \sim 0.4$; $R_{fT-4} \sim 0.45$). Iodine moves as a sharply defined zone: $R_f = 0.55$. The development time amounts to about 60-70 min. Stahl and Pfeifle [412], as well as Massoglia and Rosa [247], determined the R_f-values of some iodoamino acids on silica gel layers (Tables 26 and 27).

Table 26. Solvents[1] for the chromatography of iodoamino acids.

Number	Solvent		Author
54[2]	95% acetic acid/benzene/xylene .	3 : 1 : 1	Schneider and
55[2]	Phenol/acetone/1 N NaOH (gas phase with ammonia sat.)	2 : 7 : 1	Schneider [366]
56[3]	Butanol/dibutylether/glacial acetic acid/water	7 : 13 : 6 : 14	Hollingsworth et al. [162]
57[3]	tert.-butanol/2 N ammonia chloroform	376 : 70 : 60	
58[4]	t-amylalcohol/ammonia[5]		Patterson and Clements [315]
59[6]	Ethylacetate/isopropanol/ 25% ammonia	55 : 35 : 20	Stahl and Pfeifle [412]
60[6]	Acetone/isopropanol/ 25% ammonia	2 : 2 : 1	
61[6]	Isopropanol/25% ammonia	4 : 1	
62[6]	Ethylacetate/methanol/ diethylamine	5 : 4 : 2	
63[7]	Phenol/water	75 : 15	Massoglaia and Rosa [247]
64[7]	Butanol/glacial acetic acid/water	10 : 1 : 1	
65[7]	Butanol/methanol/20% ammonia	4 : 1 : 1	
32[7]	*Compare Table 19*		

[1] Compare also solvent number 32 of Table 19.

[2] Silica Gel G (2.5 g) + 8 ml water are applied on a 15 × 15 cm plate. Drying at 105 °C.

[3] Cellulose powder (30 g) + 140 ml water; drying in air; plates stored in dessicator.

[4] MN-300-cellulose (7.5 g) + 20 ml 0.8% (g/v) starch solution after homogenization + 30 ml starch solution. Layer thickness: 0.25 mm. Before application of the samples: washing run with the solvent.

[5] Ammonia (5 ml) + methanol to 100 ml; 20 ml of this solution + 100 ml 5-amylalcohol + 100 ml water. Shake well, separate phases. The filter paper placed into the separating chamber is impregnated with the aqueous phase.

[6] Merck Silica Gel HF$_{254}$ (25 g) + 65 ml water containing 0.5 g Merck Amylum solubile p.a. are homogenized and applied in a thickness of 0.25 mm. The plates are dried at 110 °C for 30 min. and stored over blue gel.

[7] Silica Gel G; 20 min. drying at 110 °C.

Table 27. R_f-values of iodoamino acids.

| Substance | Abbr. | \multicolumn{8}{c}{*100xR_f in solvent number*} |
		59[1]	*60[1]*	*61[1]*	*62[1]*	*63[2]*	*64[2]*	*65[2]*	*32[1]*
Monoiodotyrosine	MIT	07 (10)	20	23	08	52	41	54	64
Diiodotyrosine	DIT	07	20	23	08	77	56	45	75
Diiodothyronine	T—2	31	42	38	13	—	—	—	—
Triiodothyronine	T—3	30	39	33	17	—	—	—	—
Thyroxine	T—4	22	35	30	12	—	—	—	—
Tyrosine	TYR	—	—	—	—	40	23	26	36
Iodide	—	33	48	47	40	14	70	27	28

[1] R_f-values on Silica Gel HF$_{254}$/starch layers (compare Table 26) according to Stahl and Pfeifle [412]. Development distance: 15 cm.

[2] R_f-values on Silica Gel G layers (compare Table 26) according to Massoglia and Rosa [247].

Hollingsworth et al. [162] prefer cellulose layers and solvents numbers 56 and 57. Fig. 66 shows the separation of the iodoamino acids + tyrosine. T-3 and T-2 could not be separated here; for their separation, compare Stahl and Pfeifle (Table 27). According to Patterson and Clements [315], cellulose layers and solvent number 58 are also suited for the separation of some iodoamino acids. The sequence is DIT (start), iodide ($R_{T-2} \sim 0.33$), T-4 ($R_{T-2} \sim 0.54$), T-3 ($R_{T-2} \sim 0.83$), and T-2.

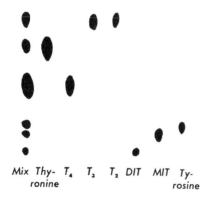

Fig. 66. Separation of the iodoamino acids on cellulose layers with solvent No. 57. From: Hollingsworth et al. [162].

Mix Thy- T$_4$ T$_3$ T$_2$ DIT MIT Ty-
ronine   rosine

Berger et al. [26] showed that MIT, DIT, and T-4 can be separated on Dowex 1 × 2 (OH$^-$) layers. However, if the I^{131}-labeled compounds are investigated in the presence of I^{131}, it can be found that T-4 and I^{131} are both at the starting point. As a result, the activity measurement of I^{131}-T-4 is disturbed by I^{131}. This unpleasant phenomenon induced the French group to search for new methods in thin-layer chromatography. Several layers are to lie side by side on one plate. The first layer is now to retain those components of a mixture which are not separated from the other components on the second layer. Berger et al. [26] use a modification of the Stahl applicator described on p. 4 for the preparation of multiple layers; a partial impregnation of a layer, as described by Honegger [167], might also be suited for the purpose. A further development of the Berger method is represented in the gradient technique of Stahl [404] (p. 28).

In the present case, I^{131} could be separated from T-4 on an AgCl layer.

2. Detection of Iodoamino Acids

The iodoamino acids can be visualized with Ninhydrin reagent (p. 90). However, it must be pointed out that the required quantities of substance are relatively high compared to iodine-free amino acids. Chromatograms of radioactively labeled derivatives can be evaluated by measuring the activity (p. 35) or by autoradiography (p. 35). All of these detection methods are non-specific, however. As a rule, several methods will be combined which have a sensitivity for nitrogen and iodine respectively.

Patterson and Clements [315] have shown that iodoamino acids can be detected with iron(III)-chloride-potassium hexacyanoferrate(III)-arsenite.

Solution I: 2.7% $FeCl_3 \cdot 6H_2O$ solution in 2 N HCl.

Solution II: 3.5% (g/v) aqueous solution of pure $K_3[Fe(CN)_6]$.

Solution III: 3.8 g a.p. arsenic trioxide are heated with 25 ml 2 N NaOH, and the homogeneous solution is cooled to 5°C. 50 ml 2 N H_2SO_4, also cooled to 5°C, was added to the solution. Finally, it is diluted to 100 ml.

Color reaction: Solutions I, II, and III are mixed in a ratio of 5 : 5 : 1 directly before use. The dried plate is sprayed with reagent solution and, after development of the color, is carefully washed with water.

Stahl and Pfeifle [412] carry out a photochemical deiodation on Merck Silica Gel HF_{254}-starch layers (Table 26); in that case, the iodo-amino acids can be recognized by the blue-violet iodine-starch reaction.

The layer is carefully dried at 100°C and, after cooling to room temperature, is sprayed with a 50% acetic acid solution. Irradiation now takes place from the top for 10 min. with four Sylvania lamps installed in reflectors from 5 cm distance (radiation at 2537 Å). The substances become visible as pale violet to brown zones. Subsequently, the layer is sprayed with 10% acetic acid up to a start of transparency. With another irradiation, the typical blue iodine-starch reaction occurs almost instantaneously. The limit of detection is about 0.5 μg.

Chapter 5

Peptides and Intermediates of Peptide Synthesis

New separation and detection methods are continuously becoming more important as a result of the elucidation of the structure and synthesis of many natural peptides. Paper chromatography and paper electrophoresis have been valuable aids for many years in laboratory peptide chemistry; recently their role has been taken over, for the most part, by thin-layer chromatography.

The chromatographic systems used for the separation of amino acids can also be used as a rule for the chromatography of peptides. In the case of high molecular weight compounds, as well as with protected di-, tri-, and oligopeptides, it is often necessary to change the chromatographic system. In such cases, the solvent will be varied first. The polyzonal technique (p. 23) is particularly suited for the determination of an optimum solvent. In paper chromatography, the choice of stationary phases was more or less invariable. In thin-layer chromatography, however, the stationary phase can be varied as desired. A suitable stationary phase is most readily found by means of gradient chromatography (p. 28).

A. THIN-LAYER CHROMATOGRAPHY OF PEPTIDES

Brenner and Pataki [49] chromatographed isomeric dipeptides and dipeptide amides [295] on Silica Gel G layers. As shown by Table 28, two of ten investigated pairs of compounds can be separated in both of the two solvents and five can be separated only in one. In two cases no separation is obtained with one of the solvents, and with the other solvent no separation is obtained in three cases. In four instances both solvents are ineffective. Wieland and Bende [438] as well as Pravda, Poduska, and Blaha [326] have been occupied with the separation of diastereomeric peptides. The R_f-values found are shown in Table 29.

Table 28. 100xR$_f$-values[1] of isomeric dipeptides on Silica Gel G layers[2] (see text p. 65). From Brenner and Pataki [49] and Pataki [295].

	$100xR_f$ in the solvent	
Dipeptide pair	A^3	B^4
H.Ala-Gly.OH / H.Gly-Ala.OH	15-16	15-14
H.Gly-Hypro.OH / H.Hypro-Gly.OH	08/13	12-13
H.Gly-Leu.OH / H.Leu-Gly.OH	37-35	27/33
H.Ala-Leu.OH / H.Leu-Ala.OH	45/37	30-31
H.Gly-Phe.OH / H.Phe-Gly.OH	36-38	32/37
H.Gly-Pro.OH / H.Pro-Gly.OH	08-09	08-07
H.Gly-Ser.OH / H.Ser-Gly.OH	12-14	12-14
H.Phe-Ala.OH / H.Ala-Phe.OH	48-45	42-40
H.Gly-Val.OH / H.Val-Gly.OH	32/27	22-24
H.Gly-Leu.NH$_2$ / H.Leu-Gly.NH$_2$	53/59[5]	25/33

[1] R$_f$-values connected with a dash can be distinguished in a parallel experiment, but their difference does not permit a separation. For the nomenclature of peptides, see [189].

[2] Other suitable solvents are numbers 6, 10, and 11 (Table 12). According to our experiences [295] with about 100 peptides, the chromatographic properties are generally acceptable up to the nonapeptides.

[3] Butanol/glacial acetic acid/water 4 : 1 : 1.

[4] Propanol/water 7 : 3.

[5] Phenol/water 4 : 1 w/w.

According to Wieland [438], all investigated diastereomeric peptides can be separated from each other on cellulose or silica gel layers. The diastereomer with two centers of identical configuration (LL or DD) always has the higher R$_f$-value. Pravda et al. [326] were able to separate two diastereomeric peptides on silica gel layers without difficulty; the benzyloxycarbonyl derivatives of two diastereomeric pairs of peptides exhibited smaller R$_f$-differences (Table 29). Finally, it should be mentioned that Riniker [340] used thin-layer chromatography to distinguish two hexa- and two octapeptides, in which only *one* amino acid (tyrosine) differs in their configuration (Table 33).

On the basis of R$_f$-value differences, most of the stereoisomers listed in Table 29 can probably be separated by preparative thin-layer chromatography. An application of thin-layer chromatography to a silica gel column failed. It seems that methanol has a partial esterifying action in the presence of silica gel and that diketopiperazines and larger molecules form from the dipeptide esters on the active surface (Wieland [438]). Sephadex column chromatography [438] is suitable for the preparative separation of the stereoisomers listed in Table 29 (columns C and D [438]).

Sephadex G-50* is swollen with pyridine/water (1 mol : 1 mol) and filled into a 100 × 1.5 cm column (charged weight : 27.5 g; charged volume: 185 cm³). Equilibration of the column is stopped when the refractive index of the discharging eluate coincides with that of the eluant. The separation can be illustrated by the following example [438]: 100 mg D- and L-Ala-L-Tyr are dissolved together in 1 ml water and

Table 29. 100xR$_f$-values of diastereomeric peptides.[1]

Peptide	*100xR$_f$ in solvent*					
	C^2	D^3	E^4	F^5	G^6	H^7
L-Met-L-Ala	47	41				
D-Met-L-Ala	34	37				
L-Ala-L-Phe	48	38				
D-Ala-L-Phe	33	28				
L-Ala-L-Tyr	55	44				
D-Ala-L-Tyr	39	35				
L-Val-L-Tyr	70	53				
D-Val-L-Tyr	52	42				
Z-L-Phe-L-Leu-OMe ...	—	—	80	87	95	82
Z-D-Phe-L-Leu-OMe ...	—	—	83	82	100	78
Z-L-Phe-L-Leu'OH	—	—	82	74	80	75
Z-D-Phe-L-Leu-OH	—	—	80	77	77	70
H-L-Phe-L-Leu-OH	—	—	63	69	52	45
H-D-Phe-L-Leu-OH ...	—	—	52	61	38	35

[1] For the nomenclature of peptides see [189] and Table 14; Z = benzyloxycarbonyl.

[2] MN-300-cellulose 20-40 g/cm²; air-dried or dried at 110°C. Solvent: pyridine/water (79 : 18 w/w). LL and DD, and DL and LD migrate together [438].

[3] Silica Gel G; methanol/water 99 : 1 [438] (layer preparation: p. 65).

[4] Silica Gel G; butanol/glacial acetic acid/water 4 : 1 : 1 [326]. (Layer preparation: p. 65.)

[5] Silica Gel G; butanol/glacial acetic acid/pyridine/water 15 : 10 : 3 : 2 [326]. (Layer preparation: p. 65.)

[6] Silica Gel G; phenol/water 3 : 1 [326]. (Layer preparation: p. 65.)

[7] Silica Gel G; isopropanol/water 5 : 1 [326]. (Layer preparation: p. 65.)

charged on the column. With an elution rate of 13.5 ml/hr with the mentioned pyridine/water mixture, 15 fractions of 11 ml each are collected after a forerun of 235 ml. With loads of 200 mg, the separation was found to be complete, with 500 mg it was extensive, and with 1000 mg it was still predominant [438].

Fig. 67 gives information on the separation: Fractions 1-7 contained only L-Ala-L-Tyr and fractions 8-15 contained D-Ala-L-Tyr (yield for L-Ala-L-Tyr 98% and for D-Ala-L-Tyr 95%).

---

* A. B. Pharmacia, Uppsala, Sweden.

Riniker and Schwyzer synthesized some analogs of hypertensin [341], including α-L-Asp¹-Val⁵-hypertensin II (A); β-L-Asp¹-Val⁵-hypertensin II (B); α-D-Asp¹-Val⁵-hypertensin II (C) and β-D-Asp¹-Val⁵-hypertensin II (D). A and B, on one hand, as well as C and D, on the other, could not be separated with any of the chromatographic systems used for paper chromatography. As shown by Table 30, A and B, as well as C and D, can be easily separated on aluminum oxide layers. This differentiation could subsequently be applied to the preparative column.

The differences between the stereoisomers (A compared to C and B compared to D) are considerably less distinct, although the D-compounds have somewhat higher R_f-values in both chromatographic systems. Table 30 contains the R_f-values of two additional hypertensin analogs.

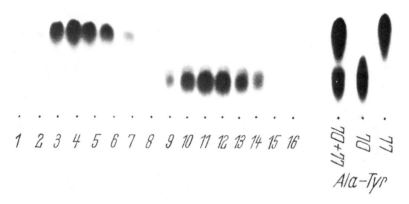

Fig. 67. Thin-layer chromatographic control of the separation of LL- and DL-Ala-Tyr on Sephadex G-50 column (pp. 99-100). From Wieland and Bende [438].

Vogler et al. [430] synthesized four isomeric cyclodecapeptides with the presumed structure of polymixin B_1 (7α, 7γ, 8α, 8γ): 40 chromatographic systems showed no differences between 7α, 7γ, and natural polymixin B_1. Paper chromatography, high-voltage electrophoresis, and IR-spectra also permitted no differentiation. The most important difference of the active synthetic compounds compared to the natural product resides in the specific rotation and particularly in the rotational dispersion of the nickel complexes at pH $= 9.3$. The fact that the two synthetic products (7α and 7γ) could not be distinguished from each other shows that the possibilities of separating very similar complex molecules are not unlimited.

Table 30. R_f-values of α-Asp- and β-Asp-peptides on aluminum oxide D5[1] layers. From Riniker and Schwyzer [341].

Peptide	$100 \times R_f$ in solvent	
	J[2]	K[3]
α-L-Asp[1]-Val[5]-Hypertensin II	34.5	39.5
β-L-Asp[1]-Val[5]-Hypertensin II	22.0	11.0
α-D-Asp[1]-Val[5]-Hypertensin II	37.0	41.7
β-D-Asp[1]-Val[5]-Hypertensin II	23.4	13.5
α-L-Asp(NH$_2$)[1]-Val[5]-Hypertensin II ...	41.4	58.0
Desamino-Val[5]-Hypertensin II	38.4	48.1

[1] Camag aluminum oxide D5 (20 g) + 50 ml water are applied on glass plates in the usual manner.

[2] sec.-Butanol/3% ammonia 100 : 44.

[3] Chloroform/methanol/17% ammonia 20 : 20 : 9.

[4] Abbreviations according to Table 14. Nomenclature, see [189].

B. CONTROL OF PEPTIDE SYNTHESES

Thin-layer chromatography can give extremely fast orienting information to the synthesist concerning a reaction process. As a rule, the distinction of the initial materials, intermediate and end products presents no difficulties since free peptides and their protected derivatives, for example, differ in polarity (Tables 31, 32, and 33). If necessary, differentiation is facilitated by the combination of several color reactions (p. 107). According to the comments on p. 99, it is often possible to separate stereoisomeric and structure-isomeric compounds. Under certain conditions, such differentiations permit the detection

Table 31. R_f-values of amino acids and of benzyoxycarbonyl (Z)-amino acids on Silica Gel G layers (see text, p. 65). From Pataki [301].

Substance	$100 \times R_f$ in ethanol/water 7 : 3	
	Free amino acid	Z-amino acid
Alanine	66.5	39
Aspartic acid	54.0	46.5
Arginine	49	03
Glutamic acid	58.5	55
Glycine	66	35
Methionine	70	51
Phenylalanine	71.5	55
Proline	67	28
Serine	67.5	40
Tyrosine	72	57
Valine	69.5	46.5

of racemization or the control of rearrangement reactions. Table 31 gives information on the separation possibilities of Z-amino acids and amino acids [301]; Table 32 shows the R_f-values of Z-amino acids, Z-peptides, and Z-peptide esters, as well as of amino acids and peptides with a free NH_2-group and the migration rate of amino acid ester hydrochlorides [106].

Table 32. R_f-values of benzyloxycarbonyl (Z)-amino acids, Z-peptides, Z-peptide esters, amino acids, amino acid ester hydrochlorides, and peptides on Silica Gel G; abbreviations, compare [106]. (Layer preparation, see text p. 65; otherwise, compare Table 14). From Ehrhardt and Cramer [106].

Compound	$100 \times R_f$ in the solvent		
	M^1	N^2	O^3
Z-Gly-OH	81	76	64
Z-Digly-OH	66	65	56
Z-Trigly-OH	57	56	50
Z-Tetragly-OH	51	43	45
Z-Gly-Gly-OEt	82	73	74
Z-Gly-Gly-Gly-OEt	76	67	69
Z-DL-Ala-Gly-OEt	81	75	77
Z-DL-Ala-OH	77	77	61
Z-DL-Diala-OH	72	72	59
Z-Gly-DL-Ala-OH	68	68	56
Z-DL-Ala-Gly-OH	68	68	57
Z-Gly-DL-Ala-Gly-OH ...	61	59	54
Z-DL-Phe'Gly-OEt	86	83	76
Z-Gly-DL-Phe-OEt	84	79	76
Z-Gly-Gly-DL-Phe-OEt ..	81	75	76
Z-Gly-DL-Phe-Gly-OEt ..	83	78	74
Z-DL-Ala-DL-Phe-OH ...	78	78	61
Z-Gly-Phe-OH	72	76	65
Z-Gly-L-Ile-OH	75	73	67
Z-Gly-L-Leu-OH	74	74	65
Z-Gly-Gly-L-Leu-OH	71	69	65
Z-Gly-L-Glu-OH	72	71	62
Z-DL-Phe	74	76	75
H-Gly-Gly-OH	17	25	15
H-DL-Ala-Gly-OH	21	30	22
H-Gly-L-Leu-OH	43	44	41
Gly	22	29	21
DL-Ala	27	34	27
DL-Phe	45	42	46
HCl-DL-Phe-OEt	59	52	65
HCl-Gly-OEt	43	42	53

[1] Butanol/acetone/glacial acetic acid/ammonia (conc. ammonia : water 1 : 4)/ water 9 : 3 : 2 : 2 : 4.

[2] Butanol/glacial acetic acid/ammonia (as above) 11 : 6 : 3.

[3] Butanol/glacial acetic acid/water/pyridine 15 : 3 : 12 : 10.

Z-peptide esters move near the front, Z-peptides move somewhat farther back, and free amino acids and peptides move more closely
to the start; amino acid ester hydrochlorides move approximately in the
middle.

Rittel [344] synthesized a number of intermediates in order to form
a nonadecapeptide with a corticotrophic activity. TLC with the use of
solvents of low polarity proved to be particularly suited for a purity
control of the protected peptides. Excellent separations were obtained
without the need to convert the substances into a detectable form by
cleavage of the protective group, as is necessary in paper chromatography. The N-protected derivatives of higher peptides, moreover, can
be easily separated from the unprotected compounds. The R_f-values of
peptides and different intermediates are compiled in Table 33, showing
the possibilities of separating the free and protected compounds.

Schellenberg [362] found that the esters of N-substituted amino acids,
di- and tripeptides have characteristic R_f-values on Silica Gel G layers
in solvents consisting of chloroform/acetone (9 : 1; 8 : 2) and cyclo-
hexane/glacial acetic acid (1 : 1). Chloroform/methanol (9 : 1), for
example, is suited for the chromatography of higher acyl-oligopeptide
esters as well as acyl-di- and tripeptide esters with polar groups in the
side chain (e.g., histidine derivatives). In that solvent, the unprotected
compounds remain at the starting point.

It would be beyond the scope of this chapter to discuss all studies
in which thin-layer chromatography was used for the purpose of con-
trolling syntheses. As an example, we can mention the control of the
total synthesis of β-corticotrophin (ACTH) according to Schwyzer and
Sieber [381]. Parts of the ACTH-molecule, starting from the C-terminal
amino acid, were built up by a step-wise addition of protected amino
acid esters. The crude protected nonatriacontapeptide, which was purified in counter-current distribution, was obtained from β^{1-10}-corticotro-
phine and β^{11-39}-corticotrophine with the aid of the dicyclohexylcarbodiimide method. The control of the purification process is shown in Fig.
68a. The purified protected ACTH-molecule was then treated with trifluoroacetic acid in order to remove the protective groups. The compound formed proved to be pure according to thin-layer chromatography (Fig. 68b); it is identical to natural ACTH (Fig. 68b).

Curtius [77] reported on the use of thin-layer chromatography for
the analysis of cysteine and cystine derivatives. Compounds with a
free SH-group, which are oxidized for the most part during the long
migration time in paper chromatograms, could be chromatographed on
thin-layers (the silica gel must be free from iron).

Other data on solvents and R_f-values of intermediate products of
syntheses and peptides, as well as other applications of thin-layer chro-

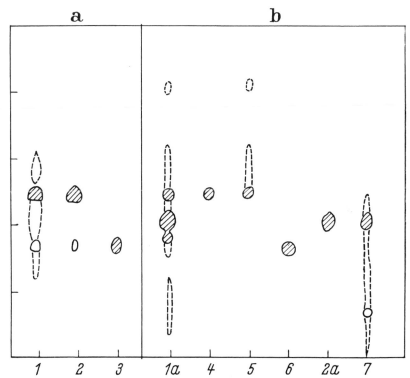

Fig. 68. Thin-layer chromatography of ACTH-peptides (synthesis control); (1) crude protected nonatriacontapeptide (ACTH); (2) protected nonatriacontapeptide after purification in counter-current distribution; (3) sulfoxide of the protected nonatriacontapeptide, (1a) crude triacontapeptide after F_3CCOOH treatment, (4) β^{1-10}-corticotropin, (5) β^{1-10}-corticotropin treated with dicyclohexylcarbodiimide, (6) β^{11-39} corticotropin, (2a) synth. ACTH [(2) after F_3COOH-treatment], (7) natural ACTH.

a) Thin-layer chromatography on aluminum oxide layers; ethylacetate/pyridine/glacial acetic acid/water 60 : 20 : 6 : 11 (layer preparation, see Table 30).

b) Thin-layer chromatography on Silica Gel G layers with butanol/pyridine/glacial acetic acid/water 30 : 20 : 6 : 24 (layer preparation, p. 65). From: Schwyzer and Sieber [381].

Table 33. $100xR_f$-values[1] of peptides and of intermediates of the syntheses on Silica Gel G layers (see text, p. 65 and [189]). Abbreviations, see [189]: BOC = butyloxycarbonyl; PZ = p-phenylazocarbobenzoxy; t-Bu = tert.-butyl; Z = Cbo.

Compound	$100xR_f$ in the solvents[2]		Author
Z-ProPro-Gly-benzylester	75(P)	76(U)	Wünsch
H-Pro-Pro-Gly-OH	14(P)	24(U)	et al.
Di-Z-Lys-nitrophenylester	80(P)	42(S)	[450]
N^α-Z-N^ϵ-BOC-Lys-nitrophenylester	80(P)	75(S)	
Di-Z-Lys-Pro-Pro-Gly-OH	64(P)	59(U)	

Table 33—Continued

Compound	$100xR_f$ in the solvents[2]		Author
N^α-Z-N^ϵ-BOC-Lys-Pro-Pro-Gly-OH	64(P)	13(W)	Wünsch
Z-Ser-Pro-benzylester	77(P)	59(T)	et al.
H-Ser-Pro-OH	17(P)	30(U)	[450]
BOC-Phe-OH	74(P)	60(T)	
BOC-Phe-nitrophenylester	81(P)	75(S)	
BOC-Phe-Ser-Pro-OH	66(P)	32(S)	
N^ϵ-Z-Lys-benzylester.HCl	69(P)	65(U)	
BOC-Phe-N^ϵ-Z-Lys-benzylester	87(P)	62(T)	
Phe-N^ϵ-Z-Lys-benzylester	27(P)	15(T)	
Phe-Ser-Pro-Phe-N^ϵ-Z-Lys-benzylester	68(P)	70(V)	
N^ϵ-BOC-Lys-Pro-Pro-Gly-OH	31(P)		
N^α-Z-N^ϵ-BOC-Lys-N^ϵ-BOC-Lys-Pro-Pro-Gly-OH	68(P)	64(U)	
Lys^{1,9}-Bradykinin	29(R)	38(U)	
Lys^{2,10}-Kallidin	21(R)	29(U)	
Z-Val-Tyr-Pro-OtBu	84(X)	81(Y)	Riniker
Z-Val-Tyr-Pro-OH	45(X)	75(Y)	[340]
H-Val-Tyr-Pro-OH	30(X)	46(Y)	
H-Val-Tyr-Val-His-Pro-Phe-OCH$_3$	76(X)	22(Y)	
H-Val-Tyr-Val-His-Pro-Phe-OH	47(X)	18(Y)	
H-Val-D-Tyr-Val-His-Pro-Phe-OH	51(X)	24(Y)	
H-Asp-Arg-Val-Tyr-Val-His-Pro-Phe-OH (= Hypertensin II)	21(X)	03(Y)	
H-Asp-Arg-Val-D-Tyr-Val-His-Pro-Phe-OH ..	26(X)	05(Y)	
Z-Glu(OtBu)-His-Phe-Arg(NO$_2$)-Try-Gly-OH	53(X)	53(Y)	
Z-Glu-His-Phe-Arg(NO$_2$)-Try-Gly-OH	24(X)	42(Y)	
Z-Val-Tyr-Val-His-Pro-Phe-OCH$_3$	77(Z)	—	
H-Val-Tyr-Val-His-Pro-Phe-OCH$_3$	57(Z)	—	
PZ-Lys(BOC)-Pro-Val-Gly-Lys(BOC)-Lys(BOC)-Arg(NO$_2$)-Arg(NO$_2$)-Pro-OCH$_3$	75(Z)	—	Kappeler and
PZ-Lys(BOC)-Pro-Val-Gly-Lys(BOC)-Lys(BOC)-Arg(NO$_2$)-Arg(NO$_2$)-Pro-OH	30(Z)	—	Schwyzer [189]
PZ-Lys(BOC)-Pro-Val-Gly-Lys(BOC)-Lys(BOC)-Arg(NO$_2$)-Arg(NO$_2$)-Pro-Val-Lys(BOC)-Val-Tyr-Pro-OtBu	65(Z)		

[1] The R_f-values have been derived from several studies; they serve solely for the demonstration of the chromatographing possibilities and for the evaluation of the R_f-changes with the introduction and cleavage of the protective group.

[2] P = Butanol/glacial acetic acid/water 3 : 1 : 1; R = butanol/glacial acetic acid/water 1 : 1 : 1; S = hexane/ethylacetate/glacial acetic acid 20 : 10 : 1; T = heptane/tert.-butanol/glacial acetic acid 3 : 2 : 1; U = butanol/glacial acetic acid/water/pyridine 30 : 6 : 24 : 20; V = tert.-amylalcohol/isopropanol/water 20 : 8 : 11; W = chloroform/methanol 9 : 1; X = sec.-butanol/3% ammonia (100 : 44); Y = n-butanol/glacial acetic acid/water 100 + 10 + about 30 (upper phase); Z = dioxane/water 3 : 1.

matography in the field of amino acid and peptide synthesis, can be found in Schwyzer et al. [144, 342, 375-380], Boissonas et al. [35, 146, 147, 174, 181], Schröder et al. [370-373], Vogler et al. [322, 419-421,

431, 432], Brenner et al. [40, 201], Rudinger et al. [145, 349], Wieland et al. [93], and others [136, 156, 157, 275, 418, 436, 455].

C. DETECTION OF PEPTIDES AND PEPTIDE DERIVATIVES

The color reactions described on p. 90 can be used for the detection of peptides. The Ninhydrin reaction is not as sensitive for the higher peptides; in cyclic peptides it even fails, if they contain no NH_2-group in the side chain. The chlorine/tolidine reaction is suitable for the coloring of free amino acids (p. 93) and peptides as well as for protected derivatives. Furthermore, Morin, chromic sulfuric acid, and iodine/starch can be used for the detection of protected peptides.

Often it is advisable to combine different color reactions. For example, the plate is sprayed with Ninhydrin (free amino acids and peptides), the spots are marked, and subsequently the chlorine/tolidine reaction is carried out (detection of protected derivatives). Finally, the layer is treated with an aggressive reagent (for example, chromic sulfuric acid). In order to permit observation of fluorescence quenching in UV-light, layers containing zinc silicate should be used (p. 163). Finally, reference should still be made to the nondestructive detection of amino acids (p. 92), which probably is also suited for the detection of peptides. This procedure is of great practical importance especially together with the "thin-layer fingerprint" technique (p. 121).

1. Chlorine/Tolidine Reaction

a) *Procedure According to Brenner et al.* [46, 299]. Approximately equal volumes (20-50 ml each) of 1.5% $KMnO_4$ solution and 10% HCl solution are mixed in a photographic developer tray of suitable size, and subsequently a grid of glass rods (feet of 2-3 cm height) is placed into the tray. The plate with the substances to be detected is briefly held over boiling water in order to moisten the layer and is subsequently placed on the grid. The tray is covered with a large glass plate (reaction period 15-20 min.). If bottled chlorine is available, it is simpler to fill a tank chamber (pp. 13, 14) with chlorine gas and to place the moistened plate (see above) into this chlorine atmosphere for 5-10 min. After completed chlorination, the plate is aired for 2-3 min. A corner of the chromatogram is sprayed first; if the background turns blue, the layer needs to be aired for a longer time. Sensitivity: 0.5 µg dipeptide or 0.5 µg Z-amino acid.

Spray reagent: 80 mg o-tolidine + 15 ml glacial acetic acid + 0.5 g KI are brought to 250 ml with dist. water.

b) *Procedure According to von Arx and Neher* [5]. Solution I: 2% aqueous sodium hypochlorite solution.

Solution II: Equal volumes of a saturated solution of o-tolidine in 2% acetic acid and a 0.85% aqueous solution of KI are mixed before use.

The chromatogram is lightly sprayed with solution I and is allowed to stand at room temperature for 1-1½ hrs. (some solvent vapors are detrimental to chlorination). Subsequently, the plate is sprayed uniformly with solution II.

c) *Procedure According to Barrolier* [13, 11]. The plate is moistened as described in a) and is treated with chlorine. Reaction solution: 100 ml 0.32% o-tolidine solution in 1 N acetic acid + 1.5 g Na_2WO_4 $2H_2O$ in 10 ml water. A white, coarse, flaky precipitate forms first which is brought into solution by the addition of 115 ml acetone. A slight turbidity remains at the start but is of no significance for the color reaction; after some time, a small quantity settles on the bottom and the solution becomes clear. Dissolved in 100 ml of this solution is 200 mg KI. The reagent solution is stable in a brown bottle.

After completion of chlorination, the plate is aired (test as in a)) and is sprayed with reagent solution. Amino acids and peptides form blue spots with a greenish tinge; it must be noted, however, that the layer must be completely penetrated by the reagent solution since otherwise an off-color of red-brown forms, particularly if the spots are very intense. In contrast to a) and b), the color is stable.

2. Other Color Reactions

a) *Iodine/Starch Reaction*. The chromatogram is placed in a concentrated iodine atmosphere for 5 min. and, after removal of the excess of iodine, is sprayed with a 1% starch solution. Blue spots form as a result; if too much iodine has remained on the layer (test a blank spot), the background also turns blue.

b) *Detection with Morin*. N-protected amino acids and peptide derivatives can be detected with morin [362]. The dried plate is sprayed with an 0.05% methanolic morin solution and heated to 100°C for 2 min. Yellowish-green fluorescence or dark absorption spots on a green fluorescing background can be recognized in UV-light. The limit of detection is about 2 μg [362].

c) *Detection with Chromic Sulfuric Acid*. The dried plate (10 min.) at 120-150°C) is sprayed with a saturated solution of $K_2Cr_2O_7$ in concentrated sulfuric acid. After heating, dark grey spots form. The limit of detection for Z-compounds is about 3 μg and for unprotected amino compounds about 1 μg [106].

d) *Detection of Cysteine and Cystine Derivatives*. Compounds without a free SH-group are identified with the usual spray reagents—chlorine/tolidine, for example (p. 107). For the detection of substances with free SH-group, the following procedure is recommended [77]. The chromatograms are dried briefly (air drying is sufficient in the case of butanol/glacial acetic acid/water) and are sprayed with a 1% sodium nitroprusside solution in water. When the plate is now briefly held over concentrated ammonia solution, red spots form which are not particularly stable and which must therefore be marked immediately.

e) *Detection of Hydrazides, Aminodiacylhydrazines, and Diacylhydrazines.* For the detection of the peptide hydrazides which are protected at the amino group [372], the layer is sprayed with a 1 : 1 solution of 1% $FeCl_3$ in 2 N acetic acid and 1% K_3 [$Fe(CN)_6$] in water. Blue spots form instantaneously.

Brenner and Hofer [40] use the detection methods listed in Table 34 for the detection of hydrazides, aminodiacylhydrazines, and diacylhydrazines.

Table 34. Methods of detection for hydrazides, aminodiacylhydrazines, and diacylhydrazines. From Brenner and Hofer [40].

Detection	Hydrazides	Aminodiacyl-hydrazines	Diacyl-hydrazines
Ninhydrin (p. 90)	—	yellow, red, violet	—
Chlorine/tolidine (p. 107)	blue	blue	blue
0.2% $FeCl_3$ in 50% methanol	—	red, violet	red, violet
Tollens[1]	brown	—	—
Methyl red/NaOBr[2]	red	—	—

[1] 0.5 ml 0.1 N $AgNO_3$ + 0.5 ml 2 N NaOH + 2 N NH_4OH dissolved and filled to 40 ml with water.

[2] (a) 0.05% methyl red in 2 N H_2SO_4; (b) 0.1 N NaOBr.

Application of Thin-Layer Chromatography in the Sequential Analysis of Proteins and Peptides

Chapter 6

Thin-Layer Chromatographic Analysis of Protein and Peptide Hydrolysates

Numerous methods have been developed for the determination of the amino acid sequence of a protein or peptide (for a summary see Bailey [6], Smyth and Elliot [397]). Proteins and peptides must be present in their pure form; their separation from lower molecular weight impurities usually presents no difficulties. In addition to earlier methods, such as dialysis or counter-current distribution, increasing importance is being attributed to Sephadex gel filtration (A. B. Pharmacia).

Sephadex is produced by cross-linking, linear macromolecules of dextrose, and it consists of a three-dimensional network of polysaccharide chains. Sephadex swells considerably in water or electrolyte solutions and forms a multipore gel. Molecules which are larger than these pores are not retained and are the first to be eluted from the column. In this manner, it is possible to fractionate on the basis of different molecular weights (Flodin [115], Determann [92], and Gelotte [135]). It is of great practical significance that the solutions are hardly diluted. Gel filtration is used most frequently for desalination and for the separation of low molecular-weight contaminants.

The separation of lower peptides from amino acids, which is often difficult, can also be accomplished by this method [92, 115, 135].

The first data concerning the composition of a protein or peptide are furnished by a total hydrolysis and end-group determination. Enzymatic or chemical cleavage results in smaller fragments, which can be separated by ion exchange chromatography, electrophoresis-chromatography, or Sephadex gel filtration. Orienting information on ion exchange and electrophoresis chromatography can be obtained from Bailey [6], Smyth and Elliot [397]; gel filtration is described by Determann [92] and Gelotte [135]; electrophoresis-chromatography on thin-layers is described on p. 118.

The amino acid sequence of the total molecule can be determined from the partial sequences resulting from the hydrolysis, end-group determination, and degradation of the peptides isolated from the partial hydrolysate.

A. AMINO ACIDS IN TOTAL HYDROLYSATES

A complete cleavage of all peptide bonds can be obtained by an acid, alkaline, and enzymatic hydrolysis or by the action of strong acid ion exchangers. Aside from enzymatic methods, certain changes and, under some conditions, even the decomposition of some amino acids must be expected with all hydrolysis methods. Asparagine and glutamine can be liberated in an undeteriorated form only by enzymatic means. Data on enzymatic hydrolysis have been reported by Hill and Schmid [159], Tower et al. [425], Barry [16], and Nomoto et al. [278]; it must be noted, however, that cleavage is not always complete. A qualitative detection of amino acids in total hydrolysates presents no difficulties (p. 116). The "tracer" method of Beale and Whitehead [19], described on pp. 61-62, is particularly suited for quantitative amino acid determinations. The accuracy amounts to about ±3%, so that it corresponds to ion exchange chromatography.

In order to obtain a reliable picture of the amino acid composition, it is advisable to use several hydrolysis methods side by side.

1. Acid Hydrolysis

Before the HCl hydrolysis of a protein or peptide, it is necessary to remove salts and carbohydrates since serine, threonine, cystine, and tyrosine are otherwise decomposed for the most part. On the one hand, methionine is converted into methionine sulfoxide and other Ninhydrin-positive compounds [321], and on the other hand methionine, homocysteic acid, and homocystine [117] form from methionine sulfoxide. Tyrosine is chlorinated [264] and even brominated [357]; however, these changes can be reduced if fresh distilled hydrochloric acid is used and small quantities of easily oxidized substances (e.g., thioglycolic acid) are added to the reaction mixture. Tryptophan is completely decomposed under the conditions of acid hydrolysis and, consequently, is chromatographed after alkaline hydrolysis. Tryptophan can also be determined in the bound form (Opienska-Blauth [286]).

The hydrolysis is conducted in carefully cleaned vacuum tubes in the absence of air. The substance is treated with a 200-500-fold excess of 6 N HCl (distilled two or three times) and is frozen in the vacuum tube.

After displacement of the supernatant air by pure nitrogen, the tube is evacuated to about 1 torr and the ampule is fused (lower part cooled with dry ice). Hydrolysis takes place at $110 \pm 1°C$ for 18-72 hrs. in a thermostat.

The removal of hydrochloric acid by allowing the product to stand over NaOH in an evacuated desiccator results in losses of serine and threonine and, therefore, cannot be recommended [397]. It is preferable to remove the HCl in a rotary evaporator.

Even under these improved working conditions, some serine and threonine losses are unavoidable. The original content of hydroxyamino acids can be determined approximately in quantitative determinations by comparing 24- and 72-hr. hydrolysates. Leucine, isoleucine and valine require about 72 hrs. of hydrolysis if they are combined in peptide linkage. Cystine is decomposed more or less and should, therefore, be determined after performic acid oxidation in the form of cysteic acid [258, 277]. In this reaction, methionine is converted into methionine sulfone, while histidine and threonine are partially decomposed. Finally, it should be mentioned that pyrrolidone carboxylic acid is formed from glutamic acid in an equilibrium reaction (ratio of 98 : 2).

Milder conditions must be used for the oxidative cleavage of disulfide bridges. The method of Hirs [160] is suited for this purpose. Tryptophan is partially decomposed, cystine converts into cysteic acid, and methionine converts into methionine sulfone. Compounds containing tryptophan after performic acid oxidation can, therefore, be used only to a limited extent for further sequential analysis.

2. Alkaline Hydrolysis

As mentioned above, tryptophan decomposes during the HCl hydrolysis. If this compound is to be detected in chromatography, the cleavage must be carried out with bases. If α-aminobutyric acid is to be reliably detected, an alkaline hydrolysis should also be used, since α-aminobutyric acid is formed from threonine in the course of an acid hydrolysis according to Heyns and Walter [158a] and according to Brieskorn and Glasz [52, 138]. During hydrolysis with $Ba(OH)_2$, cystine, cysteine, serine, and threonine are decomposed for the most part.

Five to ten mg of sample with 1 ml water + 65 mg $Ba(OH)_2 \cdot 8H_2O$ in a fused tube are heated to 125-130°C for 24 hrs. The cooled reaction mixture is adjusted to pH = 6 with 2 N H_2SO_4 and is boiled and centrifuged. The barium sulfate is washed with a small amount of distilled water. The solution and wash water are combined, evaporated to dryness, and the residue is dissolved in 0.5-1 ml water and 0.1 N HCl respectively.

3. Hydrolysis with Ion Exchangers

It has been known for some time that the peptide bond can be quantitatively cleaved by means of strong acid cationic exchangers [218, 276, 316, 249]. Hydrolysis experiments with Dowex 50(H) have shown that aspartic acid, serine, and threonine are liberated very rapidly and valine and isoleucine, on the other hand, only slowly compared to HCl hydrolysis [316]. Indoles are decomposed for the most part and glutamic acid is partly converted into pyrollidone carboxylic acid. The cleavage of cystine and cysteic acid peptides is incomplete. Most disadvantages are extensively alleviated if the reaction is carried out in 70-90% ethanol in the absence of air [324]; under these conditions, the formation of humine is also completely prevented.

According to Pöhm [324], 0.05-2 g peptide with 1 g Amberlite IR-112 (H) per milliequivalent amide nitrogen in 3-10 ml 80% ethanol are fused into an ampule under nitrogen and heated to 90-95°C for 6-10 hrs. After cooling, the amino acids are eluted from the resin with a 10% ammonia solution. The method was found useful for the cleavage of ergot alkaloids with a peptide structure and for the hydrolysis of di- and tripeptides. The expected amino acids, such as tryptophan and phenylalanine, could be found quantitatively even in the presence of carbohydrates and fats.

4. Applications

Thin-layer chromatography is an excellent method for the qualitative analysis of peptide and protein hydrolysates even if some amino acids are present in a large excess [45, 109]. It is advisable to oxidize routinely with performic acid before the chromatography (p. 115) in order to convert cysteine and cystine into the more stable cysteic acid. If the oxidation is omitted, cysteine and cystine form several secondary spots. Methionine forms methionine sulfone under the conditions of performic acid oxidation. This does not interfere either with the two-dimensional (Fig. 61a) or with the continuous flow chromatogram (Fig. 60). Larger quantities of hydrochloric acid interfere with chromatography. Consequently, the hydrolysate is evaporated several times to dryness in vacuum with a repeated addition of water. Fig. 69a shows the amino acids in a HCl hydrolysate of elastin [45, 109].

This should be compared with Table 35 in which the quantitative amino acid composition is shown. Fig. 69a and Table 35 are in very good agreement if the limits of detection of individual amino acids on silica gel layers are taken into consideration (Table 25). It is particularly remarkable to note the considerable salt tolerance. The elastin hydrolysate (Fig. 69a) could be chromatographed in the presence of approximately 4 equivalents Na^+ per equivalent of amino acid [45].

Behrens and Neu [20, 271] detected 16 Ninhydrin-positive substances in the HCl hydrolysate of the eosinophilic fraction of horse blood and were able to identify 15 of these components (Fig. 69b). All expected amino acids could be detected in the hydrolysate of the B-chain of insulin with the use of the same solvent combination [45]. Hofmann [161] chromatographed proteins on hydroxylapatite layers; in order to verify that the Ninhydrin-positive zones were actually proteins, the spots were eluted from an unsprayed chromatogram and were chromatographed after hydrolysis as described above. Gorchein [139] was able to detect ornithine in the lipids of Rhodopseudomonas spheroids after HCl hydrolysis. Identification was made by simultaneous chromatography with 5-C^{14}-ornithine on MN-300-cellulose layers and by ion exchange chromatography. Offer [283] used proteolysis and column chromatographic fractionation, as well as TLC, in order to demonstrate that myosine is an N-acetylated protein (chromatography of the degradation peptides and their hydrolysis products on Silica Gel H layers with the solvents

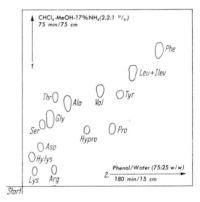

Fig. 69a. Hydrolysate of 14.6 mg prepared dried elastin (ligamentum nuchae, calf, 3 months) dissolved in 2 ml 0.2 N NaCl-HCl (pH ~ 2). The optimum load (1 μl) was determined in a series test. The composition of this solution (quantitative ion exchange chromatography) can be found in Table 35. Layer: Silica Gel G. From Fahmy et al. [45, 109].

1. Phenylalanine
2. Leucine + isoleucine
3. Proline
4. Tyrosine
5. Valine
6. Histidine
7. Alanine
8. Threonine
9. Glycine
10. Serine
11. Arginine
12. Glutamic acid
13. Lysine
14. Aspartic acid
15. ?

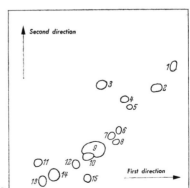

Fig. 69b. Thin-layer chromatogram of the hydrolysate of the eosinophilic substance [20, 271]. Experimental conditions as in Fig. 69a. Separation of the leucines takes place as in Fig. 60. From Neu [271].

of butanol/chloroform/formic acid 5 : 5 : 1 and isopropylalcohol/ 0.880 ammonia/water 4 : 1 : 2 as well as phenol/formic acid/water 15 : 1 : 4).

Kündig and Neukom [213] separated the glycoproteins from the water-soluble mucoids of wheat flour on DEAE-cellulose columns; after degradation with pronease P, one of the glycoprotein fractions furnished numerous amino acids which could be separated and identified in chro-

Table 35. Comparison between quantitative ion exchange chromatography and thin-layer chromatography. Note the quantitative ratio of about 25 : 1 between lysine and hydroxylysine (comapre Fig. 69a). From Fahmy et al. [45, 109].

Amino acid	$\mu g/1\ \mu l$	Visibility in the thin-layer chromatogram
Alanine	0.7	+
Arginine	0.1	+
Aspartic acid	0.1	+
Glutamic acid	0.4	−
Glycine	1.1	+
Histidine	0.03	−
Hydroxylysine	0.04	+
Hydroxyproline	0.4	+
Isoleucine	0.1	+
Leucine	0.4	+
Lysine	0.1	+
Methionine	0.02	−
Phenylalanine	0.2	+
Proline	0.6	+
Serine	0.1	+
Threonine	0.1	+
Tyrosine	0.1	+
Valine	0.6	+

matograms as shown in Fig. 59b. Kaufmann et al. [192] reported on the thin-layer chromatographic detection of the hydrolysis products of phosphatides. Silica Gel G with 96% ethanol/7% ammonia 1 : 2 as the solvent proved to be particularly suited for the separation of these substances. Different color reactions must be combined to detect these compounds which are chemically very different. In the upper part of the chromatogram, two substances could be found which had similar R_f-values as serine and threonine. Since the hydrolysis was carried out by reflux boiling for 24 hrs. with 6 N HCl, considerable losses of amino acids must be anticipated, however (p. 114).

Electrophoresis chromatography (p. 85) can also be used for the detection of amino acids in protein hydrolysates. Nybom [281] used this

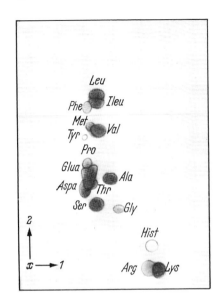

Fig. 70. Thin-layer electrophoresis-chromatography of a casein hydrolysate according to Nybom [281]. (1) Electrophoresis (compare Table 23, No. II) 15 min.; (2) chromatography (propanol/pyridine/water 5 : 1 : 2), 2 hrs. Aspa = aspartic acid; Glua = glutamic acid.

method to determine the composition of the hydrolysate of casein. The result is shown in Fig. 70.

In the examples given up to this point, the free amino acids were separated by chromatography or electrochromatography. In view of a quantitative determination (p. 56), it is often of advantage to convert the hydrolysates with dinitrofluorobenzene or with phenylisothiocyanate and to chromatograph the amino acid in the dinitrophenylated form or as thiohydantoins. The quantitative conversion of an amino acid mixture into the dinitrophenyl (DNP) derivatives and the phenylthiohydantoins is described on pp. 128 and 151.

Keller-Schierlein and Deer [197] determined and isolated larger quantities of DNP-glycine, DNP-ornithine and DNP-serine from the hydrolysate of desferri-ferrichrysine which had been converted with dinitrofluorobenzene (DNFB).

The direct hydrolysis of the sideramines with HCl leads to confusing results. Catalytic hydrogenation results in the desferri-compounds which can be cleaved with HCl or HI. Results with a good reproducibility are obtained particularly by the HI-hydrolysis. The DNP-amino acids can be isolated from the hydrolysate (p. 130), followed by chromatography (p. 131). In this connection, it should be noted that the sequence analysis of these cyclopeptides presents considerable difficulties, since only amino acids can be determined after partial hydrolysis [197].

Ishii and Witkop [178] used thin-layer and gas chromatography for the determination of the amino acid composition and configuration in gramicidin A.

Gramicidin A is hydrolyzed with glacial acetic acid/hydrochloric acid. One part each of the hydrolysate is treated with D-amino acid oxidase and with L-amino acid oxidase, respectively. The enzymes are incubated with water in a control experiment; after completion of the reaction, the mixture is converted with DNFB, and subsequently the DNP-amino acids are isolated. The latter are used directly for thin-layer chromatography, while they are converted into their methyl esters by means of diazomethane for the purpose of gas chromatography.

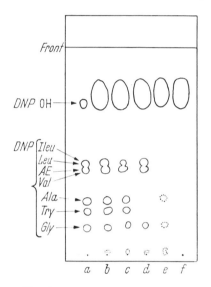

Fig. 71. Thin-layer chromatogram of dinitrophenylated amino acids from the hydrolysate of gramicidin A according to Ishii and Witkop [178]. Layer: Silica Gel G; AE = ethanolamine; solvent: chloroform/methanol/glacial acetic acid (95 : 5 : 1).

(a) Standard mixture; (b) hydrolysate; (c) hydrolysate incubated with D-amino acid oxidase; (d) hydrolysate incubated with L-amino-acid oxidase; (e) enzyme blank incubated with D-amino acid oxidase; (f) enzyme blank incubated with L-amino acid oxidase.

Fig. 71 shows the DNP-amino acids from the hydrolysate of gramicidin A with and without treatment with amino acid oxidases. Tryptophan and alanine were digested completely or nearly so by L-amino acid oxidase; consequently, they must have the L-configuration. Since valine, leucine, and isoleucine can be separated only incompletely in one-dimensional chromatograms, gas chromatography must be used in order to determine the configuration of these compounds. If necessary, the thin-layer chromatographic separation can be improved by eluting the spots and rechromatographing the DNP-amino acids in a second solvent. In a two-dimensional development, the separation of these substances presents no difficulties. Gas and thin-layer chromatography must be combined in any case, since DNP-tryptophan methyl ester cannot be determined with the gas chromatographic method used [178]. Gas chromatography offers the great advantage of permitting a direct quantitative evaluation. As shown on p. 56, a quantitative analysis by thin-layer chromatography is also relatively simple.

B. THIN-LAYER FINGERPRINT TECHNIQUE

Proteins and peptides can be split into smaller fractions under the action of acids or enzymes. A partial enzymatic degradation generally is to be preferred to HCl hydrolysis, since the latter may change the amino acid sequence. Summary reports on partial hydrolysis have been given by Sanger [356], Smyth and Elliot [397], Bailey [6], and Zuber [457].

The proteolysis product must be fractionated, for example, with ion exchange chromatography [6, 356, 397]. Another possibility is offered by two-dimensional electrophoretic-chromatographic separation on paper or thin layers. This method, known as the "fingerprint technique," was introduced by Ingram and was subsequently expanded by Anfinsen et al. [191]. Ingram [177] and Anfinsen [191] separated peptides from partial hydrolysates on paper. Wieland and Georgopoulos [440] showed that the application of the fingerprint technique to thin silica gel layers resulted in numerous advantages, such as a shorter time of analysis and savings in material. Silica Gel S layers serve for the preparation of "fingerprints"; the two-dimensional development takes place either by chromatography or electrophoresis chromatography.

For the preparation of five plates (20 × 20 cm), 25 g Macherey and Nagel Silica Gel S are stirred into 70 ml distilled water at boiling temperature. The plates are coated with the hot suspension (thickness of 0.375 mm). After 20 min. in air, they are heated in a vertical position to 110°C for 30 min. The cooled plates are ready for use. If necessary, they are stored in a closed desiccator (without drying agent).

According to Wieland [440], development is carried out with chloroform/methanol/25% ammonia (2 : 2 : 1) in the first dimension and, after drying (30 min. at 110°C), with pyridine/glacial acetic acid/butanol/water (40 : 14 : 68 : 25) in the second dimension. After drying once more (40 min. at 110°C), development is repeated with the second solvent in the second direction. A fairly useful peptide separation can already be obtained purely by chromatography [440].

Electrophoresis chromatography on Silica Gel S layers is particularly suitable for the preparation of fingerprints [440, 441]. Fig. 72 shows the tryptic peptides from ribonuclease (a) and lactate dyhydrogenase I(b), both after performic acid oxidation.

Electrophoresis is carried out with the Pherograph (Frankfurt) (compare also pp. 33-34) after the plate has been placed on the cooled surface (internal temperature −5°C) at about 40 mA; i.e., about 1000 V at both ends. The sample is applied at a distance of 2 cm from the layer

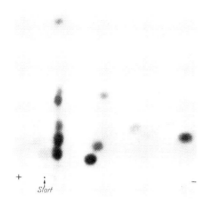

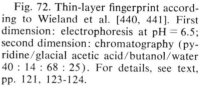

+   i
    Start   −

Fig. 72. Thin-layer fingerprint according to Wieland et al. [440, 441]. First
dimension: electrophoresis at pH = 6.5;
second dimension: chromatography (py-
ridine/glacial acetic acid/butanol/water
40 : 14 : 68 : 25). For details, see text,
pp. 121, 123-124.

LDH−I
Ninhydrin

a) Tryptic peptides from performic
acid-oxidized ribonuclease.

b) Tryptic peptides from performic
acid-oxidized LDH-isozyme I (heart).

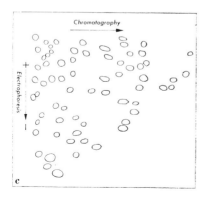

Fig. 72c. Thin-layer fingerprint of the
tryptic peptides from myosin according
to Ritschard [343]. Silica Gel G; first dimension: chromatography (chloroform/
methanol/34% ammonia 2 : 2 : 1), 1
hr.; second dimension: electrophoresis
at 980 V, 30 mA, 1 hr.; for details, compare text, p. 124.

Fig. 72d. Thin-layer fingerprint of tryptic peptides of human hemoglobin according to Faselle et al. [110]. Sephadex G-25 fine, equilibrated and developed in the first dimension with 0.02 M phosphate buffer (2½ hrs.); second dimension: electrophoresis at 460 V and 25 mA (7 hrs.).

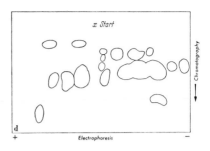

Fig. 72e. Fingerprint of the tryptic peptides of horse hemoglobin according to Stegemann and Lerch [414]; Silica Gel H (compare text); first dimension: electrophoresis at 950 V, 32 mA, 40 min.; for rest, see text pp. 124-125.

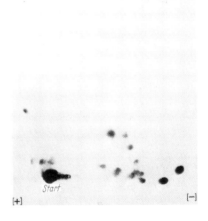

edge and 7 cm from the side of the anode (approximately 250-300 μg sample in 0.01 ml solution). Spot diameter about 3-4 mm. After evaporation of the solvent, a droplet of an aqueous 1% ammonium picrate solution is added. After drying, the plate is carefully and not too heavily sprayed with buffer (100 ml pyridine + 10 ml glacial acetic acid + 890 ml water; pH = 6.5) and is placed with the uncoated side on the cooled surface. Before applying the electrodes, the excess of buffer must be removed by application of a filter paper and rolling exclusive of the point of sample application. The electrophoresis is terminated when the yellow picrate spot has migrated for a distance of 5 cm (about 40 min.). After drying (30 min. at 110°C), development proceeds in the second dimension with pyridine/glacial acetic acid/butanol/water (40 : 14 : 68 : 25). In the case of ribonuclease, it is of advantage to repeat the chromatography after another drying period (40 min. at 110°C).

Using the above method, Wieland et al. [441] were able to confirm the hybrid nature of the isozymes of lactate dehydrogenase: The "finger-prints" of the partial hydrolysates of LDH-I and V showed, on the one hand, that these two compounds have a different primary structure and, on the other, that the peptide pattern of LDH-II was identical to that of a mixture of I and V (3 : 1).

Ritschard [343] used Silica Gel G layers (20 g Silica Gel G + 80 ml water for five 20 × 20 cm plates; air-dried) and performed the chromatography in the first and electrophoresis in the second dimension. Fig. 72c shows the tryptic peptides of myosine. The sample (0.05-0.5 mg; spot diameter 4 mm) is applied at the left top (Fig. 72c) and is developed with chloroform/methanol/34% ammonia (2 : 2 : 1). After drying (10 min. at 110°C), the cooled plate is carefully sprayed with buffer solution (pyridine/glacial acetic acid/water 1 : 10 : 489). Electrophoresis takes place at 950-1000 V and 30 mA in the second dimension (development time 1 hr.). This method permits the preparation of about 8 peptide charts per day. Ritschard [343] emphasizes the importance of the method for a purity control of peptides isolated by column chromatography and for the characterization of partial hydrolysates.

Determann [442], as well as Johansoon and Rymo [185], showed that peptides can be separated on Sephadex layers. Fasella et al. [110] separated the tryptic peptides of human hemoglobin on Sephadex G-25 layers. Gel filtration was carried out in the first dimension and electrophoresis in the second. A fingerprint on Sephadex layers is shown in Fig. 72d.

Sephadex G-25 fine (A. B. Pharmacia) is suspended in 0.02 M phosphate buffer (pH = 6.8, 50-100 ml buffer per g of Sephadex). The mixture is stirred for 30 min., the gel is then allowed to settle, and the liquid is decanted. This pretreatment is repeated 5-6 times. The gel should be equilibrated with the buffer for about 48 hrs. Fasella et al. [110] prepared layers of 1 mm thickness by suspending the equilibrated gel in the double quantity of buffer. In our opinion, the layer preparation with an applicator described by Morris [259] is more suitable for the preparation of uniform layers than that of Fasella et al. [110] (without applicator). The above-mentioned buffer is used for development in the first dimension. Fasella et al. [110] use a type of descending development (approximately 20° inclination of the plate); according to Determann [442], it is also possible to develop in the ascending direction. Continuous flow chromatography in the BN chamber probably is particularly suited for chromatography (horizontal continuous flow development, p. 20). Electrophoresis in the second dimension takes place at 460 V and 25 mA (7 hrs.).

Stegemann and Lerch [414] used Silica Gel H layers of 0.25 mm thickness (30 g Merck Silica Gel H + 75 ml pyridine/glacial acetic acid/water 20 : 9.5 : 970); electrophoresis was performed in the first dimen-

sion (pyridine/glacial acetic acid/water 3 : 12 : 485, 40 min., pH =
3.9, 950 V, 32 mA); and chromatography was carried out in the second
dimension (butanol/glacial acetic acid/water 4 : 1 : 1, 2½ hrs.). Fig.
72e shows the fingerprint of the tryptic peptides of horse hemoglobin.

Glaesmer et al. [464] reported that peptides may be separated on a
micropreparative scale with the use of 0.4 mm layers. The compounds,
eluted with 10% acetic acid, can be used directly for sequential analysis.
For degradation of peptides *without* elution from the layer, cf. p. 155.

Chapter 7

N-(2,4-Dinitrophenyl)-Amino Acids

The dinitrophenyl method of Sanger [355] is suitable for the detection of the N-terminal amino acid of a protein or peptide. As early as 1923, Abderhalden [1] used 2,4-dinitrochlorobenzene for the detection of free amino groups in silk fibroin. Since the reactivity of this compound was very low, Sanger [355] replaced it by 2,4-dinitrofluorobenzene (DNFB). Proteins and peptides react with DNFB with the formation of dinitrophenyl derivatives (DNP) (compare Scheme II, page 213).

The conversion of the free amino acids with this reagent is of considerable practical significance for their determination in biological specimens, because the DNP-amino acids can easily be separated from the interfering salts (p. 182).

A. DINITROPHENYLATION

The reaction of amino acids, peptides, and proteins with DNFB is evident from Scheme II (page 213); it is carried out in alkaline solution. In dissolved form, the DNP-amino acids obtained are particularly sensitive to light [2, 194, 325, 351-353]. Pollara and von Korff [325] found that some DNP-amino acids are decarboxylated under the influence of light and converted into the corresponding DNP-alkylamines. Russel [351, 353] observed that the DNP-derivatives of most amino acids are decomposed in NaHCO$_3$ solution under intense illumination. The induced photolysis [353] results in 4-nitro-2-nitroso-aniline. DNP-tryptophan is not modified; the DNP-derivatives of asparagine, cystine, cysteic acid, and proline are decomposed, but the decomposition product could not be identified. β-, δ- and ϵ-DNP-amino acids are photo-stable, and DNP-peptides are more stable than α-DNP-amino acids. The extraction and chromatography of DNP-amino acids, therefore, should be carried out in the absence of direct light [352]. Finally, it should be mentioned that a long reaction period and too high temperature may promote the formation of undesirable reaction products of intermediate DNP-amino acid dinitrophenyl esters (Tonge [424]).

1. Amino Acids

a) *Preparation of Dinitrophenyl-Amino Acids.** According to Levy and Chung [223], the reaction is carried out in aqueous solution. Amino acid (10 mmoles) and Na_2CO_3 (2 g anhydrous) are dissolved in 40 ml water at 40°C. This solution is treated with 10 mmoles DNFB in the form of a 10% acetone solution. Amino acids with two reactive groups require the double quantity of DNFB, and histidine requires the 2½-fold amount. The suspension is shaken vigorously for 30-90 min. at 40°C in the absence of light, while the DNFB-droplets slowly disappear. After the reaction has been completed, the product is extracted with ether in order to remove any remaining DNFB. The solution is carefully acidified with about 3 ml concentrated hydrochloric acid; the precipitated DNP-amino acids are filtered or may be extracted with ether.

According to Rao and Sober [332], the conversion takes place in ethanol solution. Amino acid and DNFB (see above) are shaken in 50% ethanol in the presence of excess sodium bicarbonate for 2-5 hrs. (for reaction conditions, see above). After the ethanol has been removed at room temperature, the product is extracted 3 times with ether. The aqueous solution is acidified with 6 N HCl, and the precipitated DNP-amino acid is washed repeatedly with small quantities of ice-cooled water.

For the purification (both methods), the precipitated DNP-amino acids are dissolved in ether; usually they already crystallize during concentration of the ether solution. Recrystallization is successful by solution in benzene to which a small amount of ethanol is added and by adding petroleum ether to the hot solution. Highly polar DNP-amino acids are recrystallized from aqueous methanol; ether-insoluble derivatives are reprecipitated by solution in dilute hydrochloric acid and neutralizing (e.g. pyridine). DNP-glutamic acid as a rule does not crystallize. It is prepared most favorably from DNP-glutamine by hydrolysis [31]. For the preparation of DNP-arginine, DNP-cysteic acid, α-mono- and di-DNP-histidine and the different mono- derivatives of cysteine, cystine, lysine, ornithine and tyrosine, see Biserte et al. [31]; for $N_{imidazolyl}$-DNP-histidine, compare Siepmann and Zahn [389], and for O-DNP-serine and O-DNP-threonine, compare Zahn and Siepmann [454].

---

* Collections can be obtained, for example, from Mann Research Laboratory, Inc., New York, or from Serva, Heidelberg.

Table 36 contains the most important water-soluble DNP-amino acids, and Table 38 shows the most important ether-soluble DNP-amino acids.

b) *Dinitrophenylation of an Amino Acid Mixture.* The conversion of an amino acid mixture (e.g. protein hydrolysate) with DNFB can be carried out in aqueous or in ethanol solution (for the conversion of free amino acids in biological specimens, see p. 182).

Dinitrophenylation in aqueous medium: According to Wallenfels [433], the dry hydrolysis residue of 2-5 mg of air-dried protein, oxidized with performic acid (p. 115) is dissolved in 2 ml CO_2-free water at room temperature with vigorous stirring. An aliquot (1.2 ml) is pipetted into a small reaction vessel with a magnetic stirrer, it is diluted with 1.8 ml CO_2-free water, 0.1 ml 3.1 N KCl is added, and the solution is heated to $40 \pm 0.1\,°C$. The pH is now adjusted to 8.9 by the addition of 0.2 N NaOH with vigorous stirring. Approximately 0.1 ml DNFB is added in a small excess in the absence of light, and the pH is held at 8.9 for 100 min. by means of an autotitrater. The reaction kinetics can be followed by automatic recording of the alkali uptake. After the reaction is terminated, the excess DNFB is removed by extracting twice with 5 ml each of peroxide-free ether. If ether with a peroxide content is used, DNP-methionine sulfone forms from DNP-methionine. The reaction mixture is acidified (0.5 ml of a solution of 1 part HCl, $d_4^{20} = 1.19 + 1$ part water) and the ether-soluble DNP-amino acids are extracted by treating five times with 4 ml peroxide-free ether. The extracts are combined and brought to 25 ml with ether. For the quantitative chromatography (p. 56), 1 ml is removed, is concentrated somewhat, and is quantitatively applied on the layer. During the reaction, 2,4-dinitrophenol (DNP-OH) forms from DNFB. If this compound is present in larger quantities, it may mask a few of the DNP-amino acids under certain conditions (p. 138). In such cases, it is advisable to remove the dinitrophenol (p. 130). The aqueous phase contains α-mono-DNP arginine, α-mono-DNP-histine, DNP-cysteic acid, and a part of di-DNP-histidine (compare Table 36). After removal of dissolved ether, the aqueous phase is filled to 10 ml with CO_2-free water. For the chromatography, 0.5 ml is removed, is evaporated to dryness in vacuum, is taken up in a minimum amount of acetone (possibly acetone/6 N HCl 2:25), and is quantitatively applied on the layer. Biserte et al. [31] prefer the following procedure. The solution of the water-soluble DNP-amino acids is extracted repeatedly with sec.-butanol/ethylacetate (1 : 1), the extracts are combined, evaporated to dryness, and taken up in acetone as described above. The acetone solution contains the mentioned water-soluble DNP-amino acids.

Dinitrophenylation in alcohol solution: According to Lucas et al. [232], the dry hydrolysis residue of 2-5 mg protein that had been dried over P_2O_5 and possibly oxidized with performic acid is dissolved in 5 ml water. This solution is treated with 100 mg $NaHCO_3$ and a solution of 100 mg DNFB in 8 ml ethanol. The single-phase mixture obtained is allowed to stand for 3 hrs. at room temperature in darkness. After termination of the reaction, the ethanol is evaporated in vacuum; the temperature should not exceed 40°C. Now 50 mg $NaHCO_3$ are added and the product is extracted three times with ether. The aqueous phase is acidified with concentrated HCl to such an extent that the HCl-concentration amounts to about 1-2 M. Subsequently, it is extracted five times with ether; the extracts are combined, are evaporated to dryness in vacuum, and are dissolved in acetone or ethylacetate. Under these conditions, di-DNP-histidine, DNP-arginine, and DNP-cysteic acid remain in the aqueous phase. These compounds are chromatographed directly as described on p. 131 or are extracted with sec.-butanol/ethylacetate (1 : 1).

Although histidine in alcohol solution is reported to furnish only di-DNP-histidine according to the literature [31], it was found by Lucas et al. [232] that a reaction of 16 hrs. is necessary for a complete conversion. The American authors recommend that two reactions be conducted simultaneously; one for the determination of all amino acids without histidine, and the other for the determination of histidine.

2. Peptides

a) *Conversion with Dinitrofluorobenzene.* Dinitrophenylation according to Sanger and Thompson [358]; A solution of 0.2 μ mole peptide in 0.1% trimethylamine is treated with 10μl DNFB dissolved in 0.2 ml ethanol. After 2 hrs. in darkness, a few drops of water and trimethylamine solution are added, followed by extracting three times with fresh ether. The solution is evaporated in vacuum to dryness.

Dinitrophenylation according to Lockhart and Abraham [231]: 50-150 μg peptide are dissolved in 0.1 ml 1.5% (g/v) trimethylammonium carbonate solution (pH ~ 9.3), 0.2 ml of a 5% alcohol solution of DNFB is added, and the mixture is allowed to stand in darkness for 2½ hrs.; the ethanol is then evaporated in vacuum, the product is treated with 0.24 ml trimethylammonium carbonate solution and 1 ml ether, followed by mixing with a vibromixer, centrifuging in order to separate the phases, separation of the ether (discard) and evaporation of the aqueous solution in vacuum to dryness.

b) *Total Hydrolysis of a Dinitrophenyl Peptide.* The DNP-peptide is taken up in 0.1 ml 6 N HCl and is heated for 9 hrs. under nitrogen to 105°C in a bomb tube (concerning the losses during hydrolysis, see below). The hydrolysate is diluted with 2 vols. water and extracted with ether (ether-soluble DNP-amino acids). The water-soluble DNP-amino acids are applied directly or are first extracted with sec.-butanol/ethylacetate, are taken up in acetone (p. 131) and chromatographed.

3. Polypeptides and Proteins

a) *Conversion with Dinitrofluorobenzene.* According to Levy and Li [229], 0.2 μ mole substance is dissolved in 30 ml 0.05 N aqueous KCl at 40°C, the pH is adjusted to 8 (addition of 0.05 N KOH by an autotitrater), about 0.1 ml DNFB is added, and the mixture is stirred vigorously in darkness, while the pH and temperature are maintained constant. The reaction is terminated when no further base is consumed. The solution is extracted three times with ether and then is acidified. The precipitated DNP-compounds are centrifuged, are washed with water, acetone and ether, and dried over P_2O_5.

b) *Total Hydrolysis of a Dinitrophenyl Protein.* For a qualitative end-group determination, the DNP-protein is hydrolized with the 100-fold quantity of double-distilled 5.7 N hydrochloric acid at 105°C for 16 hrs. (in a bomb tube under nitrogen). Under these conditions, some DNP-amino acids, such as the DNP-derivatives of glycine, proline, tyrosine and cystine (the latter is best converted to cysteic acid), are decomposed for the most part [31]. According to Fittkau et al. [114], DNP-glycine and DNP-proline are not significantly modified in control experiments. DNP-glycine, DNP-proline, and other acid-sensitive DNP-derivatives are decomposed to a considerably smaller degree if the hydrolysis is carried out with Dowex-50 (H) (Steven [416]). If the end-groups are to be determined quantitatively, several hydrolyses should be carried out side by side (variation of the time, HCl concentration and reagent). The hydrolysate is diluted to such an extent that the HCl-concentration amounts to about 1 N. Extraction proceeds three times with peroxide-free ether and five times with ethyl-acetate for the isolation of di-DNP-histidine. The extracts are washed with 0.1 N HCl. The ether and ethylacetate extracts are combined and the aqueous phase is combined with the wash water. *Ether-soluble DNP amino acids:* Larger quantities of dinitrophenol may be unfavorable for the separation under certain conditions (p. 138). DNP-OH can be removed by sublimation [31] (where methionine, as well as cysteine, are partially lost) or by adsorption on aluminum oxide [427] or on silica gel [417].

Silica gel (5 g, 100 mesh Mallinkrodt) is carefully mixed with 2.5 ml 0.067 M Na_2HPO_4, suspended in chloroform, and filled into a 30 × 1 cm column. The ether-soluble DNP-compounds are most suitably charged on the column in chloroform. Elution is carried out first with chloroform, saturated with 0.067 M Na_2HPO_4; dinitrophenol and dinitranaline are only partly ionized in this alkaline medium and are therefore eluted. On the other hand, the ionized DNP-amino acids are retained on the column. They are eluted with chloroform containing 1% glacial acetic acid. The eluate is finally concentrated and taken up in acetone or ethylacetate.

Water-soluble DNP acids: The following compounds can be present in this fraction: DNP-arginine, DNP-cysteic acid, mono-DNP-derivatives of cysteine, cystine, histidine, lysine, ornithine, and tyrosine, as well as a small amount of di-DNP-histidine. These compounds can be directly identified by chromatography in spite of the presence of free amino acids (care is necessary in the Ninhydrin reaction). The aqueous solution is evaporated several times to dryness with a repeated addition of water and is taken up in 0.5 N HCl, in glacial acetic acid or in acetone/ glacial acetic acid (about 1 ml per 10 μ mole protein; load about 1 μl; care should be taken that the acid solvent is completely evaporated before chromatography). Otherwise, the water-soluble DNP amino acids can be extracted with sec.-butanol/ethylacetate (1 : 1) (p. 128). Another possibility is offered by a selective adsorption of the DNP-amino acids (except DNP-cysteic acid) on Hyflo-Super-Cel-Talc [31].

c) *Partial Hydrolysis of a Dinitrophenyl Protein.* The acid or enzymatic partial hydrolysis of a DNP-protein is of considerable practical importance in sequential analysis. In chromatograms or "fingerprints," the N-terminal fractions are recognized by their yellow color, while other than N-terminal fractions are colorless (exception: non-N-terminal lysine peptides contain ϵ-DNP groups and, consequently, are also yellow).

Peptides containing ϵ-DNP-lysine can easily be separated from the free peptides and amino acids [31]. Fractionation of the α- and ϵ-DNP-peptides is often difficult, at least when extraction is used. For other data, see Keil [193].

B. THIN-LAYER CHROMATOGRAPHY OF WATER-SOLUBLE DINITROPHENYL AMINO ACIDS

The few DNP-amino acids can be chromatographed most favorably with solvents which consist of a mixture of an alcohol (n-propanol or n-butanol) and ammonia solution. DNP-arginine, DNP-cysteic acid, mono-DNP-cystine, α-DNP-histidine, di-DNP-histidine, ϵ-DNP-lysine, and O-DNP-tyrosine can be chromatographed on Silica Gel G layers with n-propanol/34% ammonia (7 : 3) as solvent. The R_f-values are listed in Table 36 [44].

ϵ-DNP-lysine and DNP-arginine are not separated in this chromatographic system; nevertheless, these compounds can be determined side by side due to their differences in the Ninhydrin reaction (Table 36).

DNP-cysteic acid and mono-DNP-cystine probably never occur together. α-DNP-lysine ($R_f = 0.1$) and δ-DNP-lysine ($R_f = 0.2$) can be separated on Silica Gel G layers with ethylacetate/pyridine/glacial acetic acid/water (60 : 20 : 6 : 11) [378].

For the chromatography of water-soluble DNP-amino acids formed during the dinitrophenylation of urine-, blood-, and sperm-amino acids, etc., see Table 50.

Table 36. Identification of the water-soluble dinitrophenylamino acids[1] on Silica Gel G layers (see text p. 65 and Table 14). From Brenner et al. [44].

Substance	$100 \times R_f$[2]	Color	Color with Ninhydrin
Mono-DNP-(Cys)$_2$	29	yellow	brown
DNP-CySO$_3$H	29	yellow	yellow
α-DNP-Arg	43	yellow	yellow
ϵ-DNP-Lys	44	yellow	brown
O-DNP-Tyr	49	colorless	violet
α-DNP-His[3]	57	yellow	yellow
Di-DNP-His[3]	65	yellow	yellow

[1] For the chromatography of water-soluble DNP amino acids from biological specimens, see Table 50.

[2] Average values from 6 single determinations each; solvent: n-propanol/34% ammonia (7 : 3); ascending technique; development distance: 10 cm; load: 0.5-1 μg.

[3] The im-DNP-histidine also belongs into the group of water-soluble DNP amino acids. According to Zahn and Pfannmüller [453], this compound should not be detectable due to its instability in hydrolysates of DNP proteins and DNP peptides; it was, therefore, not investigated in greater detail by Brenner et al. [44]. More recent studies of Zahn [389] have shown, however, that im-DNP-histidine can be prepared by a simple method and can also be detected in DNP-protein and DNP-peptide hydrolysates.

The sample (1 μg) is applied in, e.g., 1 μl glacial acetic acid or 0.5 N HCl. It is important to remove the excess acid from the layer prior to chromatography. For this purpose, the plate must be heated to about 60°C for about 10 min. with good ventilation after application of the solution. Munier and Sarrazin [471, 473] separate the most important water-soluble DNP-compounds on cellulose layers with the use of electrophoresis chromatography.

C. THIN-LAYER CHROMATOGRAPHY OF ETHER-SOLUBLE DINITROPHENYL AMINO ACIDS

1. Solvents and R_f-Values

According to Brenner et al. [44, 45, 434], Silica Gel G layers and solvents numbers 66-72 are suited for the chromatography of ether-soluble DNP-amino acids (Table 37).

Table 37. Solvents for the separation of ether-soluble dinitrophenyl-amino acids.

No.	Solvent		Author
66[1,2]	Toluene/pyridine/2-chloro-ethanol/0.8 N ammonia solution	100 : 30 : 60 : 60	Brenner et al. [44, 45, 434]
67[2]	Toluene/pyridine/2-chloro-ethanol/25% ammonia ..	50 : 15 : 35 : 7	
68[2]	Chloroform[8]/benzylalcohol/glacial acetic acid	70 : 30 : 3	
69[2]	Chloroform[8]/t.-amylalcohol/glacial acetic acid	70 : 30 : 3	
70[2]	Benzene/pyridine/glacial acetic acid	80 : 20 : 2	
71[2,3]	Chloroform[8]/methanol/glacial acetic acid	95 : 5 : 1	
72[2]	Chloroform[8]/methanol/glacial acetic acid	70 : 30 : 5	
73[2,3]	Chloroform[8]/methanol/glacial acetic acid	98 : 2 : 1	Keller and Pataki [196]
74[2]	n-butanol/saturated with 0.1% ammonia		Vogler et al. [432]
75[4]	1 M NaH$_2$PO$_4$ + 0.5 M Na$_2$HPO$_4$		Brandner and Virtanen [39]
76[5]	Chloroform/methanol/glacial acetic acid/petroleum ether	70 : 10 : 5 : 5-10	Fittkau et al. [114]
77[5]	Benzene/pyridine/glacial acetic acid	70 : 30 : 3	
78[6]	80-90% ethanol		Wohlleben [446]
79[6]	Dimethylformamide/glacial acetic acid/ethanol/water	5 : 10 : 20 : 30	
80[2]	Toluene/pyridine/glacial acetic acid[7]	80 : 10 : 1	Witkop [444]
81[2]	Benzene/glacial acetic acid[7] .	3 : 1	

[1] Two-phase solvent; the lower phase serves for the pretreatment of the layer (compare text).

[2] Silica Gel G (see text p. 65).

[3] Separation is highly dependent upon atmospheric humidity [196].

[4] MN-300-Cellulose (see text p. 76).

[5] Supergel (see text p. 66).

[6] Woelm polyamide.

[7] For separation of DNP amino acid methylester.

[8] Containing about 1.5% alcohol.

The use of solvent number 66 requires some special instructions. A separating chamber (pp. 13, 14) lined with filter paper is charged with the lower phase of the solvent. A heavy bent glass rod is placed on the bottom of the chamber as a support. Two plates coated with Silica Gel G are placed on the center of the support with the coated side facing out and each is allowed to lean on a wall of the chamber with its upper edge. In order to prevent the solvent from contacting the layer, the latter is divided into two parts by drawing a line parallel to the upper edge. The unit is allowed to stand overnight. This results in a large moisture absorption by the silica gel, indicated by a considerable diffusion of substances applied on the plate during the pretreatment. Chromatography on a layer pretreated in this manner probably is due to a "true" partition between two liquid phases. In any case, in plates which have been exposed to air without protection, the effect of equilibration is practically lost after just 45 min. This effect must be taken into account with the use of the plates.

If the plate is allowed to be exposed to air without protection between the pretreatment and chromatography, the R_f-values decrease after just 2 min. Consequently, after removal from the chamber, the layer must be immediately covered with a second glass plate, leaving only a strip of about 1.5 cm width exposed at the lower edge. This is where the substances are now applied. If this requires not more than about 5 min., no significant disturbance is produced. Incidentally, application can be accelerated if relatively concentrated solutions are used. *The volume applied should not exceed 1 µl.* The plate is now immediately subjected to chromatography. Development proceeds with 120 ml of the upper phase.

Table 38 gives information on the R_f-values in some of the solvents mentioned.

In addition to the R_f-values obtained with one-dimensional chromatography, those obtained in the second dimension after previous chromatography with solvent number 66 ("indirectly obtained R_f-values") are also given. They are reproducible only if the instructions for the use of solvent number 66 (see above) and those for intermediate drying* are followed exactly. Fittkau et al. [114] use Supergel layers and solvents numbers 76, 77, and 67. Wohlleben [446] uses Woelm polyamide layers and solvents numbers 78 and 79. For special separation problems, good results have been obtained with Silica Gel G layers and solvent number 74 (Vogler et al. [432]) (separation of DNP-leucine and γ-DNP-

---

* Intermediate drying M (multiple chromatography): air draft, room temperature (about 3 min.); layer turns white. Intermediate drying T (two-dimensional chromatography): 10 min in a draft, 10 min. heating to 60°C, 15 min. cooling in a draft.

Table 38. 100×R_f-values[1] and reference number of ether-soluble dinitrophenylamino acids with one-dimensional, ascending, or horizontal chromatography on Silica Gel G layers (see text p. 65). From Brenner et al. [44].

Reference Number	Compound	66² Ascending	68³ Ascending	68³ Indirect⁴	69³ (Ascend-ing) Ascending	69³ Indirect⁴	70⁵ (Horizontal⁶) Ascending²	70⁵ Horizontal⁶	70⁵ Indirect⁴	71⁵ (Horizontal⁶) Ascending²	71⁵ Horizontal⁶	71⁵ Indirect⁴
1	DNP-Ala	34	54	35	60	34	32	33	38	59	66	58
2	DNP-β-Ala	27	71	57	73	50	189	98	100	99	95	102
3	DNP-Aad[7]	—	—	—	—	—	—	—	—	—	—	—
4	DNP-Abut	46	72	44	73	42	52	52	55	79	85	75
5	DNP-γ-Abut	—	—	—	—	—	—	—	—	—	—	—
6	DNP-Acy	79	92	66	83	57	105	108	109	108	101	106
7	DNP-β-Aib	—	—	—	—	—	—	—	—	—	—	—
8	DNP-Asp(NH₂)	—	—	—	—	—	—	—	—	—	—	—
9	DNP-Asp	02	13	08	09	13	06	05	11	07	06	06
10	Di-DNP-Cys	—	—	—	—	—	—	—	—	—	—	—
11	Di-DNP-(Cys)₂	—	03	02	01	01	00	00	02	00	02	02
12	DNP-Glu(NH₂)	—	—	—	—	—	—	—	—	—	—	—
13	DNP-Glu	01	26	17	31	21	12	12	23	12	12	14
14	DNP-Gly	27	32	22	40	23	17	18	22	31	38	31
15	Di-DNP-His	53	11	09	08	04	05	04	08	12	16	14
16	Di-DNP-Hylys	—	—	—	—	—	—	—	—	—	—	—
17	DNP-Hypro	—	—	—	—	—	—	—	—	—	—	—
18	DNP-Ile	64	83	63	81	57	107	107	107	100	101	104
19	DNP-Leu	66	82	62	80	54	100	100	100	100	100	100
—	DNP-Nleu	69	82	60	80	52	86	90	88	101	100	98
20	Di-DNP-Lys	74	56	35	60	30	12	13	19	66	73	65
21	DNP-Met	55	70	39	69	38	43	43	47	72	81	74
22	DNP-MetO₂	17	—	—	—	04	03	03	02	10	10	07
23	Di-DNP-Orn	70	34	23	40	20	06	06	10	39	46	39
24	DNP-Phe	67	75	46	74	41	44	46	52	81	86	76
25	DNP-Pro	29	65	41	67	38	58	59	62	78	84	75
26	DNP-Sar	23	56	35	57	32	34	35	41	59	65	60
27	DNP-Ser	15	11	10	11	10	09	10	14	07	08	07
28	DNP-Thr	20	17	13	15	12	12	14	20	09	11	11
29	DNP-Try	65	69	38	69	31	23	25	33	54	61	49
30	Di-DNP-Tyr	76	58	35	60	30	17	16	19	57	65	57
31	DNP-Val	53	79	56	77	51	76	81	85	91	98	86
—	DNP-Nval	56	77	52	76	48	65	70	75	86	95	89
32	DNP-OH[8]	41	100	76	83	55	22	21	23	148	102	111
33	DNP-NH₂[9]	90	90	84	72	63	115	128	129	131	101	115

[1] Average values of 6 single measurements.
[2] Development distance: 15 cm.
[3] Development distance: 15 cm.
[4] After chromatography in solvent number 66 and intermediate drying (see text).
[5] R_f-values referred to DNP-leucine.
[6] Development distance of DNP-leucine: 10 cm.
[7] α-aminoadipic acid.
[8] 2,4-dinitrophenol.
[9] 2,4-dinitroaniline.

135

Table 39. 100xR$_f$values[1] of DNP-amino acid- and dipeptide-methylesters on Silica Gel G layers (cf. text p. 66) [444] (ascending).

DNP-*Methylester* from	Solvent 80	Solvent 81
Gly	30	50
Try	30	62
Ala	39	58
Val	55	67
Ile	66	70
Leu	66	72
Val-Gly	16	20
Ile-Gly	20	24
Leu-Gly	23	—
Ala-Leu	25	—
Ala-Val	23	—
Gly-Try	07	—
Val-Val	36	—
Leu-Try	25	—

[1] Abbreviations according to Table 14 and [189].

Table 40. Separation effects and development times in one-dimensional chromatography of ether-soluble dinitrophenylamino acids on Silica Gel G layers [44, 196] (see text p. 65).

Solvent Number	Separation of DNP-amino acids from Dinitro-aniline[1]	Dinitro-phenol[1]	Separation of DNP derivatives from Leu Ile Nleu Val Nval		Separation of Di-DNP-derivatives from Tyr and Lys	Development Time
66	+	−[2]	−	−	−	1 hr./15 cm
68	+	+	−	−	−	1½ hr./10 cm
69	−[6]	−[3]	−	−	−	1 hr./15 cm
70[7]	+	−[4]	+	+	−	2-3 hrs.
71[7,8]	−[5]	−	−	+	+	2-3 hrs.

[1] 2,4-dinitrophenol and 2,4-dinitroaniline are byproducts of the synthesis and acid hydrolysis of DNP-proteins and peptides.

[2] Dinitrophenol is located between DNP-Val and DNP-Ala.

[3] Dinitrophenol is located near DNP-Leu.

[4] Dinitrophenol is located between DNP-Ala and DNP-Gly.

[5] Dinitrophenol and dinitroaniline are just above DNP-Leu.

[6] Dinitroaniline is located between DNP-Phe and DNP-Met.

[7] Continuous flow chromatography in the BN chamber (see text, p. 20). DNP-Leu runs a distance of about 10 cm in 2½ hrs.

[8] The separation depends highly upon the atmospheric humidity; sometimes, solvent number 73 has better separation properties for this reason. For the separation of Di-DNP-Tyr and -Lys, ether/glacial acetic acid 95 : 5 can also be used (Brenner et al. [45]).

diaminobutyric acid) as well as MN-300-cellulose layers and solvent number 75 (Brandner and Virtanen [39]) (separation and quantitative determination of DNP-hydroxyglutamine and DNP-hydroxyglutamic acid; the R_f-values are 0.3 and 0.4).

Methylesters of DNP-amino acids and DNP-dipeptides can be chromatographed on Silica Gel G layers (solvents numbers 80 and 81) according to Witkop [444]. Table 39 contains the R_f-values found.

2. One- and Multidimensional Separation

The identification of an ether-soluble DNP-amino acid, as a rule, requires several chromatograms. Table 40 gives information on separation effects and development times for the one-dimensional chromatography in some solvents.

Three two-dimensional chromatograms serve for the separation of the most important ether-soluble derivatives [44, 45, 434]. The characteristic spot patterns are rarely changed by R_f-fluctuations if the instructions are carefully observed. Consequently, an unknown sample can be chromatographed together with a standard mixture, applying about 0.2 μg per compound of standard solution. This results in a just barely visible spot pattern, and the DNP-amino acids present in the sample are recognized by an increasing intensity of their spots (p. 21).

Method A: The mixture of ether-soluble DNP-amino acids is applied on three Silica Gel G plates. The plates are developed in the first dimension with solvent number 66 (see p. 132 for pretreatment). After intermediate drying T (the intermediate drying conditions must be precisely observed—see footnote on p. 134), each plate is chromatographed in the second dimension with solvents numbers 68 and 70, and with number 71, respectively. Fig. 73 shows the distribution of the ether-soluble DNP-amino acids.

This method is suited for the detection and determination of the N-terminal amino acid of a protein or peptide as well as for the detection and determination of amino acids in protein hydrolysates (pp. 119 ff).

Method B: The mixture of ether-soluble DNP-amino acids is separated on Silica Gel G layers in solvents numbers 67 and 68 by two-dimensional chromatography; after immediate drying M, the development is repeated in the same direction with solvent number 67, followed by an intermediate drying T (p. 134: the intermediate drying conditions must be most carefully observed). The development distance of the solvent should amount to 10-15 cm in each case (Fig. 74a). In chromatograms such as that of Fig. 74a, di-DNP-tyrosine and di-DNP-lysine never appear separated. The separating quality of di-DNP-cystine, DNP-asparagine, and DNP-glutamine, as well as of DNP-aspartic acid, DNP-glutamic

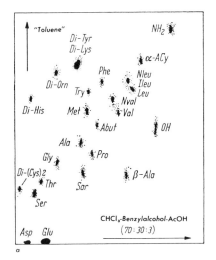

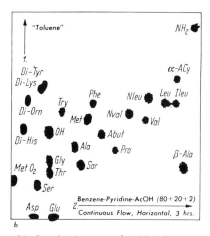

a) Load: 0.2 µg each; development distance 12 × 10 cm; first dimension: "toluene" = solvent number 66; second dimension: solvent number 68.

b) Load: 1 µg each; "development distance" 15 × 14 cm; first dimension: solvent number 66; second dimension: solvent number 70 (continuous flow, 3 hrs. in the BN chamber).

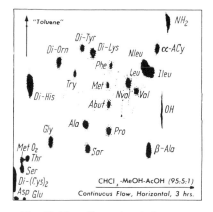

c) Load: 1µg each; "development distance": 13 × 13 cm; first dimension: solvent number 66; second dimension: solvent number 71 (continuous flow, 3 hrs., in the BN chamber). The separation in the second direction depends on atmospheric humidity. Solvent number 73 sometimes exhibits better separating properties [196]. For the separation of Di-Lys and Di-Tyr, see also Table 40. See p. 134 for intermediate drying.

Fig. 73. Two-dimensional chromatograms of a DNP-amino acid standard mixture [44]. OH = dinitrophenol; NH₂ = dinitroaniline; Di = Di-DNP-derivatives. UV-photocopy; compare text, p. 141 and Table 37. From Brenner et al. [44].

acid, and DNP-α-amino-adipic acid, and that of DNP-leucine and DNP-isoleucine depends upon the relative quantities, and, in the case of biological material, upon the possible presence of impurities. If the compounds mentioned are to be determined separately, the following procedure is used. To separate the DNP-derivatives of lysine, tyrosine, isoleucine, and leucine, development is carried out with solvent number 67 in the first dimension and, after intermediate drying T, with solvent

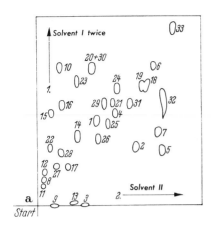

a) First dimension: I = solvent number 67 (development two times; compare text); second dimension II = solvent number 68.

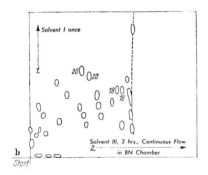

b) First dimension I = solvent number 67; second dimension III = solvent number 71 (continuous flow, 3 hrs., in the BN chamber). The separation in the second direction depends upon the atmospheric humidity. Solvent number 73 sometimes has better separating properties. For the separation of Di-DNP-Lys and Di-DNP-Tyr, see also Table 40.

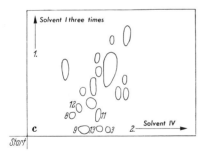

c) First dimension I = solvent number 67 (development three times, compare text); second dimension IV = solvent number 72.

Fig. 74. Ether-soluble DNP-amino acids: spot pattern (load: 0.4 µg each). The numbers refer to the reference number of Table 38. From Walz et al. [434].

number 71 in the second direction (continuous flow chromatography in the BN chamber, development time 3 hrs.) (Fig. 73c).

The successful separation of the leucines depends highly upon the relative and absolute quantities. If only one leucine/isoleucine spot forms and does not resolve into two separate spots with a decrease of the chromatogram load (for the resolution of closely adjacent spots by

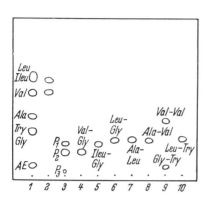

Fig. 75. Thin-layer chromatogram of DNP-amino acid and DNP-peptide methyl ester in the hydrolysate (after esterification) and in the methanolysate of DNP-seco-gramicidin A. Silica Gel G; solvent number 80 (compare also Table 39). (1) DNP-amino acid methyl ester (AE = ethanol-amine); (2) hydrolysate esterified with diazomethane; (3) methanolysate; (4) DNP-Val-Gly-methyl ester; (5) DNP-Ile-Gly-methyl ester; (6) DNP-Leu-Gly-methyl ester; (7) DNP-Ala-Leu-methyl ester; (8) DNP-Ala-Val-methyl ester; (9) DNP-Gly-Tyr-methyl ester (below) and DNP-Val-Val-methyl ester (top); (10) DNP-Leu-Try-methyl ester. From Ishii and Witkop [179].

photocopy, see p. 142), then chromatography in the second dimension is carried out with solvent number 70 (Fig. 73b). A complete separation of the DNP-derivatives of asparagine, glutamine, aspartic acid, glutamic acid, α-amino-adipic acid, and cystine is possible by chromatographing three times in the first dimension with solvent number 67 (after intermediate drying M each time) and, after intermediate drying T, in the second dimension with solvent number 72 (drying: p. 134).

The separation of the compounds mentioned can be seen in Figs. 74b and c.

Method B is suitable for the detection and determination of the N-terminal amino acid of a peptide or protein (compare method A), for the detection and determination of amino acids in protein hydrolysates (see also method A), and particularly for the qualitative and quantitative determination of amino acids in biological specimens (p. 182). Recently, studies have been published dealing with thin-layer chromatography of ether-soluble DNP-compounds on cellulose [471, 473] and polyamide [484, 485].

According to Witkop [444], the methyl esters of DNP-amino acids can be separated on Silica Gel G layers (Table 39). This method also is suited for the determination of the N-terminal amino acid [179].

If gramicidin A is treated with anhydrous hydrochloric acid in abs. methanol, a compound forms which contains all building blocks of gramicidin A and, in addition, a free amino group ("seco-gramicidin A") [179].

Preparation and hydrolysis of DNP-seco-gramicidin A: 90 mg seco-gramicidin A are dissolved in 4 ml ethanol and treated with 90 mg dinitrofluorobenzene (DNFB) and 2 ml 10% triethylamine. The mixture is allowed to stand in darkness at room temperature for 12 hrs., and then the solvent and triethylamine are evaporated. The residue is washed four times with 5 ml ether (removal of DNFB; discard the ether) and is dissolved in 3 ml ethanol (DNP-seco-gramicidin A is insoluble in ether). DNP-seco-gramicidin A is precipitated by the addition of 10 ml 0.1 M salt solution, the precipitate is washed with water, dried and dissolved in ethylacetate. For purification, the solution is applied on thick layers, developed with solvent number 71 and, finally, the DNP-seco-gramicidin A is extracted. In order to cleave the N-terminal amino acid, 3.7 g of the compound are treated with 0.8 ml glacial acetic acid and 0.8 ml 12 N HCl. Hydrolysis takes place at 110°C for 18 hrs. in the absence of air. The product is concentrated and the residue is taken up in 3.5 ml 0.5 N aqueous NaHCO₃ solution. The solution is washed three times with 2 ml ether (discard the ether), is acidified with 0.5 ml 6 N HCl and extracted three times with 2 ml ether. The ether-soluble DNP-derivatives are dissolved in a small quantity of dimethoxyethane, and the solution is treated with 1 ml diazomethane in ether. After removal of excess reagent, the residue is dissolved in 0.1 ml dimethoxyethane. For chromatography, 2-4 µl are applied on the layer (Silica Gel G, p. 65).

Fig. 75 shows the ether-soluble DNP-amino acids (as their methyl esters) from the hydrolysate of DNP-seco-gramicidin A. One of these spots is identical to DNP-valine-methyl ester (Table 39), and the other has the same R_f-value as DNP-leucine and DNP-isoleucine-methyl ester (Table 39). The latter compounds can be separated by gas chromatography [179].

Methanolysis of DNP-seco-gramicidin A: DNP-seco-gramicidin A (4 mg) is subjected to methanolysis with 2 ml methanol, half-saturated with HCl at 0°C, for 20 hrs. under nitrogen at 75°C. The reaction mixture is concentrated and the residue is distributed between 3 ml ether and 4 ml 0.1 N HCl. The aqueous layer is extracted twice with 2 ml ether and the combined ether extracts are evaporated to dryness.

In Fig. 75 we show the ether-soluble DNP-derivatives from methanolized DNP-seco-gramicidin A. Three spots can be recognized, P_1, P_2, and P_3, which have the same R_f-values as DNP-Val-Gly-methyl ester, DNP-Ile-Val-Gly-methyl ester, and DNP-Val-Gly. P_1 and P_2 can be isolated with preparative thin-layer chromatography. Their structure is confirmed by hydrolysis (DNP-Ile and DNP-Val in the ether and glycine in the aqueous phase). After P_3 is preparatively isolated on thick layers, it is converted with diazomethane and chromatographed again. Since P_1 and P_2 form from P_3 after esterification, P_3 must be a mixture of DNP-Ile-Gly and DNP-Val-Gly.

D. DETECTION METHODS

With the exception of O-DNP-tyrosine, all DNP-derivatives are yellow. Quantities of 0.1 µg (one-dimensional chromatography) or 0.5 µg (two-dimensional chromatography) can be readily recognized in transmitted daylight. The chromatograms should be permanently recorded within a few hours, since the spots gradually fade with time. The chromatograms can be copied on transparent paper with back-lighting. With some practice, even the weakest light yellow spots can be recognized on the white layer. However, it is more convenient and reliable to examine the plate in UV light. The UV-lamp of the Camag Company is particularly suited for this purpose. The DNP-amino acids are visible as intensely dark spots.

Photocopies and photography can be used for recording purposes.

1. Photocopy [44, 45, 434]

X-ray film or photocopy paper, for example, Gevaert Gevacopy paper or Agfa Brovira Normal, is placed directly with its sensitive side on the thin layer and pressed against it with a glass plate. This is followed by exposure to UV-light (360 mμ)* and development in the usual manner. The highest sensitivity is obtained with an exposure period which produces only a bright gray, but not yet a black background. With a longer exposure, the background becomes increasingly blacker; at the same time, the spots decrease in diameter. This can be used profitably in order to resolve large, partially overlapping spots into smaller separate spots. Fig. 73a, b, and c were obtained in this manner (Gevaert Gevacopy paper).

For a reliable detection of a DNP-amino acid with UV-photocopies, it is sufficient to use $2 \cdot 10^{-3}$ μmole, while 10^{-4} μmole, e.g. 0.02 μg DNP-serine, even suffice for one-dimensional chromatography.

2. Photography [434]

For the detection of very weak spots located between those of high intensity, direct photography is suitable. With photography in long-wave UV-light, very weak spots are visualized particularly well. This may be useful in cases of doubt, since very weak spots in the direct vicinity of more intense spots can often be reliably recognized in this manner.

Daylight and UV-photography are often used to record the DNP-amino acids from biological material (p. 185).

Photography in normal light: The chromatogram is illuminated from the left and right by two 8-Watt Sylvania lamps† (30 cm distance), is

---

* Uvanalys Laboratory instrument 57 US of the Bälz Company, Heilbronn, Germany.

† Draco & Co. Ltd., Zurich, Switzerland.

exposed for about $\frac{1}{15}$ sec. (Rolleicord Va with Rollinar close-up lens No. 2, F/No. 8, Agfa-Isopan IFF $\frac{13}{10}$ DIN), and development proceeds in the customary manner.

UV-photography (360 mμ): The chromatogram is illuminated from the left and right by two 8-Watt Sylvania UV-lamps (F8T5/BLB; 30 cm distance), is exposed for 5—10 sec. (Rolleicord Va with Rollinar close-up lens No. 2 and UV-filter F/No. 5.6—8; Agfa Isopan IFF $\frac{13}{10}$ DIN), and development proceeds as usual.

Chapter 8

N-(2,4-dinitro-5-aminophenyl)- and 1-dimethylaminonaphthalene-5-sulfonyl amino acids

Two reagents suited for the determination of the N-terminal amino acid in proteins and peptides are 2,4-dinitrofluoroaniline [27] and 1-dimethylaminonaphthalene-5-sulfonyl chloride [140].

Bergmann and Bentov [27] reported on the conversion of some amino acids with 2,4-dinitrofluoroaniline (DNFA). N-(2,4-dinitro-5-amino-phenyl)-amino acids (DNAP-amino acids) form in this reaction. The DNAP-amino acids exhibit a bathochromic shift (400-410 mμ) compared to the DNP-amino acids (p. 131). Since they contain an aromatic NH$_2$-group, they can be diazotized and coupled to a dye. The preparation of 1-dimethylaminonaphthalene-5-sulfonyl amino acids (DNS-amino acids) by means of 1-dimethylaminonaphthalene-5-sulfonyl chloride (DNSC) and their use for the sequence analysis of peptides have been described by Hartley et al. [140, 155]. The sensitivity of the amino acid detection with DNSC is approximately one hundred times higher than with dinitrofluorobenzene [140].

A. PREPARATION

1. Dinitroaminophenyl-Amino Acids [27]

Phenylalanine (0.8 g), NaHCO$_3$ (0.9 g), DNFA (1.5 g), and ethanol (30 ml) are heated until a clear solution is obtained. The mixture is allowed to stand for 30 min. at room temperature, is filtered, and 20 ml water is added. The solution is concentrated in vacuum and acidified with dilute hydrochloric acid. DNAP-phenylalanine can be recrystallized from methanol. The original study [27] contains additional data on the synthesis, melting point, and spectra of some DNAP-amino acids.

2. 1-Dimethylaminonaphthalene-5-Sulfonyl Amino Acids [80]

One microliter of a 10 m mole amino acid solution, 1 μl NaHCO$_3$ solution (0.1 M), and 2-5 μl DNSC-solution (10 m mole in acetone) are incubated in a capillary for 2 hrs. at room temperature. Subsequently, 5-10 vols. 2% triethylamine in 10% ethanol are added and the mixture is evaporated in vacuum to dryness. The original method of Hartley and Massey [155] is suitable for the preparation of larger quantities. The synthesis of ϵ-DNS-lysine, O-DNS-tyrosine, and im-DNS-histidine requires special procedures [155]. For the chromatography, it is best to apply 0.5 μ M of a DNS-amino acid in ethanol [80], even though considerably smaller quantities can be detected (pp. 147-148).

B. DETERMINATION OF THE N-TERMINAL AMINO ACID OF A PEPTIDE [140]

10^{-3} μM of a peptide are dissolved in 15 μl 0.1 M NaHCO$_3$, and the solution is treated with 15 μl DNSC in acetone (1 mg/1 ml). The reac-

Table 41. Solvents for the chromatography of dimethylaminonaphthalene sulfonyl amino acids on Silica Gel G layers (see text, p. 65).

No.	Solution		Author
82	Chloroform/benzylalcohol/glacial acetic acid	100 : 30 : 5	Buchanan et al. [80, 81]
83[1]	2-butanone/triethylamine/ dimethylsulfoxide/benzene ...	4 : 2 : 1 : 3	
84[2]	Benzene/pyridine/glacial acetic acid	80 : 20 : 2	
85	Chloroform/benzylalcohol/glacial acetic acid	30 : 10 : 1	
86	2-butanone/propionic acid/ water	15 : 5 : 6	
87	2-butanone/glacial acetic acid ..	13 : 5	
88	2-butanone/glacial acetic acid/ chloroform	13 : 5 : 15	
89	2-butanone/propionic acid	2 : 1	
90	2-butanone/propionic acid/water	5 : 5 : 1	
91	Methylacetate/isopropanol/conc. ammonia	45 : 35 : 20	Seiler and Wiechmann [384]
92	Chloroform/methanol/glacial acetic acid	75 : 20 : 5	
93	Chloroform/ethylacetate/ methanol/glacial acetic acid ..	30 : 50 : 20 : 1	

[1] Chromatography at 75°C.

[2] According to Brenner et al. [434], this solvent is also suited to chromatograph ether-soluble DNP-amino acids (see Fig. 73).

tion mixture is evaporated to dryness in vacuum, the residue is taken up with 20 μl 6 N HCl and heated to 105-110°C for 6-12 hrs. DNS-glycine is decomposed after 20 hrs. [80]. After removal of the acid, the DNS-amino acids can be extracted from the still acidic solution by means of organic solvents or they can be chromatographed directly. Dimethyl-aminonaphthalene sulfonic acid (DNS-OH) forms during the reaction. Although DNS-OH generally does not interfere with chromatography

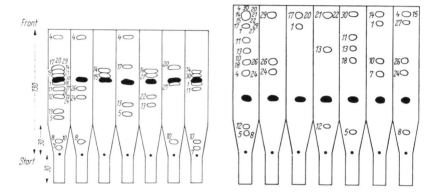

Fig. 76. One-dimensional chromatograms of DNS-amino acids on Silica Gel G layers; wedge strip development (the dimensions are given in mm) with solvent number 91 (a) and number 92 (b). The solid spot is DNS-OH. From Seiler and Wiechmann [384].

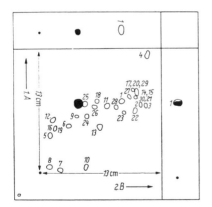 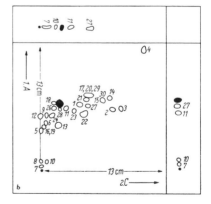

Fig. 77. Two-dimensional thin-layer chromatograms of DNS-amino acids on Silica Gel G layers.

a) First dimension: A = solvent number 91; second dimension B = solvent number 92. The solid spot is DNS-OH (see text). For the reference number, see Table 42.

b) First dimension: A = solvent number 91; second dimension C = solvent number 93. The solid spot is DNS-OH (see text). For the reference number see Table 42. From Seiler and Wiechmann [384].

(p. 146), this compound can be removed on a short Dowex 50(H)-column [140]. The solutions of DNS-amino acids must be protected from light, since they are gradually subject to photochemical degradation (Seiler and Wiechmann [384]).

C. CHROMATOGRAPHY

1. Solvents and R_f-Values

Little has been published on the thin-layer chromatography of DNAP-amino acids. According to Ratney [335], the DNAP-derivatives of alanine, diaminopimelic acid, and glutamic acid can be developed on air-dried Silica Gel G layers with n-propanol/water (7 : 3).

Silica Gel G layers and the solvents listed in Table 41 are suited for the chromatography of DNS-amino acids.

Table 42 contains the corresponding R_f-values in some solvents.

2. One- and Two-Dimensional Separation

David et al. [80, 81], as well as Seiler and Wiechmann [384], reported on the one-dimensional chromatography of DNS-amino acids. The American authors used Silica Gel G layers and solvents numbers 82, 83, and 84 for the detection of N-terminal amino acids of peptides of known amino acid composition but unknown sequence. Seiler and Wiechmann [384] prefer the wedge strip technique (p. 21), since it results in sharper separations. Although one-dimensional development is not sufficient for complete separation in any case, it is often satisfactory for the solution of many separating problems (Fig. 76).

A nearly complete separation of 30 DNS-amino acids is possible by the combination of solvents numbers 91 and 92, as well as of 91 and 93, according to Seiler and Wiechmann [384]. Two Silica Gel G plates are developed perpendicular to the direction of application with the use of solvent number 91 in the first and, after drying (10 min., 110°C), with solvents numbers 92 or 93 in the second direction. The conditions for intermediate drying must be observed very carefully since the substances otherwise move at a higher rate than the standard substances at the edge of the plate. They can even crowd together near the solvent front. The spot patterns (Fig. 77) can be easily reproduced.

For one- and two-dimensional chromatography of DNS-amino acids, cf. also references [460, 461, 469, 488].

D. METHODS OF DETECTION

For the detection of DNAP-amino acids, the plates are sprayed with a solution of N_2O_3 in toluene (10 ml toluene spread over 10 ml 6 N

Table 42. 100xR$_f$-values of DNS-amino acids with ascending chromatography on Silica Gel G layers [81] (see text, p. 76).

Ref. No.	Compound[1]	82	83[2]	84	85	86	87	88	89	90
1	DNS-Ala	48	54	47	39	88	—	60	86	—
2	DNS-β-Ala	—	—	—	—	—	—	—	—	—
3	DNS-γ-Abut	—	—	—	—	—	—	—	—	—
4	"Ammonia"	—	—	—	—	—	—	—	—	—
5	DNS-Arg	00	00	—	01	58	—	—	—	—
6	DNS-Asp(NH$_2$)	—	—	—	—	—	—	—	—	—
7	DNS-Asp	03	00	—	—	38	—	—	—	—
8	DNS-CySO$_3$H	—	00	—	—	—	—	—	—	—
9	DNS-Glu(NH$_2$)	—	—	—	—	—	—	—	—	—
10	DNS-Glu	09	00	—	05	71	—	—	—	—
11	DNS-Gly	28	23	29	22	79	88	46	77	—
12	Di-DNS-His	19	52	—	22	95	—	—	—	—
13	DNS-Hypro	—	—	—	—	—	—	—	—	—
14	DNS-Ile	—	74	73	—	—	82	—	—	—
15	DNS-Leu	73	72	73	67	97	92	70	92	94
16	ε-DNS-Lys	—	—	—	—	—	—	—	—	—
17	Di-DNS-Lys	00	02	—	00	50	—	—	—	—
—	DNS-Met	59	48	—	46	94	—	—	—	—
18	DNS-MetO$_2$	—	—	—	—	—	—	—	—	—
19	δ-DNS-Orn	—	—	—	—	—	—	—	—	—
20	Di-DNS-Orn	—	—	—	—	—	—	—	—	—
21	DNS-Phe	70	50	—	55	96	—	—	—	—
22	DNS-Pro	65	35	—	52	89	—	68	—	—
23	DNS-Sar	—	—	—	—	—	—	—	—	—
24	DNS-Ser	07	36	11	04	70	63	15	59	74
25	DNS-Tau	—	—	—	—	—	—	—	—	—
26	DNS-Thr	06	55	—	07	78	—	—	67	78
27	DNS-Try	41	35	—	33	97	—	—	—	—
28	O-DNS-Tyr	—	—	—	—	—	—	—	—	—
29	Di-DNS-Tyr	62	55	—	60	95	—	—	—	—
30	DNS-Val	70	65	66	54	97	87	67	—	—

Header note: *100xR$_f$ in solvent number*

[1] Abbreviations of the amino acids according to Table 14.
[2] Chromatography at 75°C.

HCl; after the addition of 10 ml 3 M NaNO$_2$, N$_2$O$_3$ forms and converts into the toluene phase; the toluene layer is immediately used for spraying and may not be stored), and subsequently with a solution of phenol and diethylamine (9.4 g phenol + 14.6 g diethylamine + toluene to 100 ml). The light yellow DNAP amino acids form rust-brown spots after this treatment. The limit of detection is about 10^{-3} μM [334].

10^{-4} μM of a DNS-amino acid can be detected if the moist plate is examined under UV-light (360 mμ). The intensity of fluorescence decreases extensively after drying. DNS-amino acids have a yellow and DNS-OH a green fluorescence [384].

Chapter 9

3-Phenyl-2-Thiohydantoins

PTH-amino acids (3-phenyl-2-thiohydantoins) form when proteins or peptides are treated with phenylisothiocyanate (PITC) and the condensation product is degraded in a suitable manner (Edman [102]). The formation of PTH-amino acids is illustrated in Scheme III [103, 104, 175] (p. 213).

Instead of proteins or peptides, free amino acids can also be converted with PITC. Since most PTH-amino acids are soluble in organic solvents, the amino acids can be separated from interfering impurities (e.g. salts) with this method.

A. PREPARATION OF PTH-AMINO ACIDS

The methods of Edman [102] and Cherbuliez [66], as well as the micromethod of Sjöquist [391], are suited for the preparation of PTH-amino acids. Phenylthiocarbamyl(PTC)-amino acids are formed in the first reaction step (Scheme III) and are subsequently converted into the corresponding phenylthiohydantoins in acid solution. More recent studies of Ilse and Edman [175] show that PTC-amino acids have a tendency to autoxidative desulfuration. Consequently, it is advisable to carry out the ring-closing reaction in a nitrogen atmosphere.

Sjöquist [391] as well as Ilse and Edman [175] have studied the kinetics of formation of PTH-amino acids from the PTC-derivatives. The Australian authors [175] found that practically all PTC-amino acids are converted nearly quantitatively into the corresponding phenylthiohydantoins if the reaction is carried out for 60 min. at 80°C and pH = 1. Only the yield of PTH-glycine is somewhat lower (85%). PTH-serine and PTH-threonine present some difficulties since these compounds convert into α, β-unsaturated products by a β-elimination in the side chain. Ilse and Edman [175] observed an inverse proportionality between the hydrogen ion concentration and rate of β-elimination. At

149

pH = 3.4, this secondary reaction has a rate which is approximately 7 times greater than at pH = 1. It is particularly worthy of note that these α-, β-unsaturated compounds enter a further reaction. They result in compounds with an absorption maximum of 272 mμ; presumably they are polymerization products [175] and very seldom interfere with chromatography.

Special methods have been developed for the synthesis of PTH-tryptophan [102], PTH-serine, PTH-threonine, and PTH-cystine [176, 224] as well as of PTH-Asp(NH$_2$) and PTH-Glu(NH$_2$) [331]. All PTH-amino acids are sensitive to light. Optically active derivatives are easily subject to racemization but are still sufficiently stable to permit a determination of the configuration by rotatory dispersion [99].

1. Microsynthesis According to Sjöquist [391]

A solution of 0.5 μM amino acid in 50 μl buffer (2 ml 2 N acetic acid + 1.2 ml triethylamine + distilled water to 25 ml + 25 ml acetone; pH 10.1) is treated with 50 μl PITC-solution (6 μl PITC in 1 ml acetone) and is incubated for 2½ hrs. at 25 °C. The solution is concentrated and is subsequently dried over P$_2$O$_5$. The residue is taken up in 100 μl distilled water and 200 μl glacial acetic acid saturated with HCl. The solution is allowed to stand for 6 hrs. at 25 °C, it is concentrated and the PTH-amino acid is dried over P$_2$O$_5$. For the chromatography, 0.5 μg of sample in 0.5 μl acetone or methanol is applied.

2. Synthesis According to Edman [102]

10 mM amino acid are dissolved in 25 ml water and 25 ml pyridine. Then the solution is adjusted to pH = 9 by the addition of 1 N NaOH, is heated to 40 °C, treated with 2.4 ml PITC, and maintained at pH 9 by the addition of 1 N NaOH with continuous stirring. After about 30 min., the reaction is complete and the NaOH-consumption ceases. The product is extracted repeatedly with benzene; the aqueous solution is neutralized with HCl, is concentrated somewhat, and the precipitated phenylthiocarbamyl(PTC)-amino acid is filtered. PTC-arginine and PTC-histidine require pH 7 and pH 3.5, respectively, for precipitation. The PTC-amino acid is dissolved or suspended in 30 ml 1 N HCl. It is boiled with refluxing for 2 hrs., is repeatedly evaporated with several additions of water up to dryness, and the crude compound is crystallized from glacial acetic acid/water or ethanol. See section 3 concerning the purification by sublimation.

3. Synthesis According to Cherbuliez [66]

An amino acid (0.05 mole) is dissolved in 100 ml water + 100 ml dioxane + 0.06 mole PITC; the solution is heated to 40 °C for 4 hrs., under a nitrogen stream, is cooled, extracted with 3 × 100 ml benzene (discard), 3 × 100 ml cyclohexane (discard), and the hydrocarbons are eliminated under nitrogen. The aqueous solution is slowly acidified with about 28 ml concentrated HCl (pH = 1) and is heated to 40°C for 4 hrs. under nitrogen. Most PTH-amino acids precipitate quantitatively; in the case of PTH-Ile, PTH-Leu and PTH-Phe, the solution must be concentrated a little. The PTH-amino acids are recrystallized from ethanol or are purified by sublimation (170-190°/10 Torr). With the exception of PTH-derivatives which contain free NH_2-, CO-NH_2-, COOH or heterocyclic groups, as well as of PTH-serine, all compounds investigated can be sublimed. Special methods are necessary for PTH-tyrosine and PTH-serine [66]. The original study presents data on yields, melting points, and sublimation properties. All PTH-derivatives prepared proved to be pure according to thin-layer chromatography.

4. Quantitative Conversion of an Amino Acid Mixture into PTH-Derivatives [393]

Peptide or protein (0.5-1 mg) is fused into a quartz tube with 0.3 ml double-distilled hydrochloric acid of constant boiling point (5.7 N) under nitrogen and is heated for 22 hrs. at 110°C for the purpose of hydrolysis; the cooled hydrolysate is repeatedly evaporated in vacuum to dryness with repeated water additions; the residue is treated with 250 μl buffer (2 ml 2 N acetic acid + 1.2 ml triethylamine + distilled water to 25 ml + 25 ml acetone; pH = 10.1) and with 250 μl PITC acetone solution (corresponding to 6 μl PITC); the carefully mixed sample is allowed to react in a well-closed tube for 2½ hrs. in a water bath of 25°C; the reaction mixture is freed from solvent for 15 min. with a water-jet pump and subsequently over P_2O_5 in a high vacuum. The residue (PTC-amino acids) is dissolved in 100 μl distilled water and 200 μl glacial acetic acid/saturated with HCl, and the solution is held at 25°C for 6 hrs. The solvent and acid are removed in the manner described above. The residue contains the PTH-amino acids. A parallel experiment without amino acid addition is carried out for a chromatographic comparison. Cysteine and cystine must be oxidized with performic acid before the hydrolysis (p. 115). For chromatography, the phenylthiohydantoins are dissolved in 90% acetic acid.

B. DEGRADATION OF PEPTIDES AND PROTEINS

The Edman degradation [102] is suited not only for the determination of N-terminal amino acids, but also for a step-wise degradation of peptides and proteins (Scheme III, p. 213). Although the number of degradation steps is theoretically unlimited, identification of the separated PTH-amino acids often becomes difficult after several degradation steps. For recent results, cf. [489].

The conversion of a peptide with PITC is by no means quantitative as demonstrated by Sjöquist [392]. "Artifacts" could already be found in the second step during the step-wise degradation of insulin [392]; these were the PTH-derivatives of glycine and phenylalanine, i.e. of the N-terminal amino acids of the unchanged insulin molecule. If the insulin molecule and the peptide formed are converted three times consecutively with PITC, the expected phenylthiohydantoins form exclusively. We were able to detect a part of the unconverted peptides by thin-layer chromatography after a single conversion of some peptides with PITC [298]. Therefore, it was proposed [298] to observe the formation of PTC-peptides by thin-layer chromatography (p. 153).

As indicated by Scheme III, the N-terminal amino acid is split off as thiazoline. This rapid reaction is of a non-hydrolytic nature (Edman [103]) and demands highly acid conditions. The unstable thiazoline derivative is converted into the PTH-amino acid by heating or by dilute acids. According to Edman [104], the cleaved thiazoline derivative should be separated from the residual peptide and converted into the PTH-amino acid in a separate reaction. Among the methods described in the literature, we can distinguish two groups. The cleavage of thiazoline can be carried out in an anhydrous or in an aqueous medium.

If glacial acetic acid, saturated with HCl gas, is used for the cyclization, different secondary reactions can occur [206, 397, 398]. N-terminal glutamine may convert into pyrrolidonecarboxylic acid under these conditions, i.e. the amino group is blocked. Serine or threonine rests in the molecule are acetylated quantitatively; if an O-acetylated amino acid now moves to the N-terminal position, an O → N-acyl migration takes place and the free amino group is also blocked. Finally, peptides containing aspartic acid or glutamic acid form imides and convert into the corresponding β-peptides. This new bond is not split in a further degradation step. These secondary reactions may lead to misinterpretation if only the hydrolysate of the residual peptide is analyzed ("subtractive method") and if the PTH-amino acids are not identified in addition [398]. The difficulties can be eliminated for the most part if ring-closure takes place in trifluoroacetic acid for 1 hr. at room temperature and if a purification by column chromatography is also carried out (Konigsberg and Hill [206]).

The cleavage of thiazoline can also be carried out in an aqueous medium, in solution [119, 392, 394, 67], on filter paper [118, 369], or on silica gel [439]. Although a hydrolysis cannot be ruled out in the presence of water, it has often been possible to perform several degradation steps and to identify the PTH-amino acids [154, 369, 392, 439]. This method also permits a quantitative determination of the N-terminal amino acid [107, 207, 369]. (Cf. also reference [489].)

1. Stepwise Degradation of Peptides

a) *Peptide Degradation According to Sjöquist* [*391, 392, 394*]. Peptide (0.5-5 mg) + 0.5 ml buffer (2 ml 2 N acetic acid + 1.2 ml triethylamine + water to 25 ml + 25 ml acetone; pH = 10.1) + 0.5 ml of a PITC solution in acetone (6 μl PITC in 1 ml acetone) are incubated in a closed reaction tube for 1 hr. at 40°C.

Thin-layer chromatography is used to verify the conversion (Pataki [298]). The reaction mixture is chromatographed on zinc silicate-containing Silica Gel G layers (p. 163), e.g., with n-propanol/water (7 : 3), the plate is examined in UV-light (254 mμ), the UV-positive zones are marked, and the plate is subsequently sprayed with Ninhydrin. If a more intense spot appears in the Ninhydrin reaction, it can be assumed that the conversion with PITC has been incomplete. In such cases, coupling with PITC is repeated once more or several times after removal of the byproduct (see below). Each conversion should be monitored in the manner described. The byproducts can be separated, for example, by applying bands of the reaction mixture on thicker layers and by isolating the PTC-peptide after localization in UV-light. In this manner, mono- and diphenylthiourea (MPTU and DPTU) as well as PITC are removed; the compounds mentioned are chromatographed also for control purposes. The byproduct also can be eliminated for the most part by extraction. In that case, the reaction mixture is concentrated to one-half its volume under nitrogen and washed three times with 0.5 ml benzene/ethylene chloride of 3 : 1 (the solvent is first washed with 0.1 N NaOH). The wash solution is then discarded and the aqueous phase is evaporated to dryness at 40°C under nitrogen.

The dried residue (40°C, 5-12 hrs.) is treated with 0.1 ml water and 0.2 ml glacial acetic acid saturated with HCl-gas, is allowed to stand in a closed vessel for 1 hr. at 40°C, evaporated to dryness at 40°C under nitrogen, and dried over KOH. The residue is suspended in 0.5 ml of the aqueous phase of ethylacetate/ethylene chloride/water (3 : 1 : 4), and the PTH-amino acids are extracted three times with 0.5 ml of the organic phase. The aqueous solution contains the PTH-derivatives of histidine, arginine, and cysteic acid as well as the residual peptides. PTH-histidine can be extracted almost completely at pH = 7. The organic phase contains the PTH-amino acids and possibly a small part of degraded peptide. It is concentrated to one-half volume under nitrogen and extracted three times with 0.2 ml of the organic phase mentioned above. The combined aqueous extracts are evaporated; the peptide is now ready for the second degradation step. The PTH-amino acids are dissolved in methanol, acetone, or 90% acetic acid. PTH-histidine, PTH-arginine, and PTH-cysteic acid are most suitably applied from the aqueous phase; the residual peptide does not interfere with chromatography.

b) *Peptide Degradation According to Sjöholm [390]*. As described in a), the peptide is converted with PITC (for the thin-layer chromatography of the reaction mixture, see a). In order to cleave the thiazoline, the PTC-peptide is treated with anhydrous trifluoroacetic acid (40°, 15 min.), a solution of dichloroethane/ether (0.75 + 0.9 ml) and 0.25 ml water is added, the mixture is shaken vigorously and centrifuged. The aqueous phase is extracted twice with 0.25 ml dichloroethane/ether (see above). The organic phases (thiazoline) and the aqueous phase (residual peptide) are evaporated to dryness. The thiazoline derivative is treated with 30% ethanol adjusted to pH = 1 with HCl, is heated for 75 min., extracted twice with 0.75 ml and three times with 0.5 ml dichloroethane/ethylacetate/n-heptane (4 : 4 : 3), and is evaporated to dryness. See a) for further procedures.

c) *Peptide Degradation According to Schroeder [369]*.   The aqueous solution of 0.4-1 μM peptide is applied in drops on several strips of (1 × 7 cm) Whatman No. 1 paper; each strip should be loaded with about 0.2 μM. The strips are dried for 30 min., are impregnated with 0.2 ml of a 20% PITC solution in dioxane, and are suspended in pairs in a 250 ml bottle with a screw cap over 15 ml pyridine/dioxane/water (1 : 1 : 1). The bottle is closed (opening is covered with aluminum foil before screwing on the top) and is held at 40°C for 3 hrs. The strips are dried until they are no longer translucent and are then transferred into a test tube filled with benzene. If the strips are treated with benzene without intermediate drying, some PTC-peptides may go into solution; however, if they are dried too thoroughly, the DPTU is extracted incompletely. Extraction takes place twice for 90 min. and once overnight with benzene. The strips are dried in air in a fume-free chamber (1 hr.), are allowed to react for 7 hrs. at room temperature (pressure = 100 Torr) in a desiccator (containing 15 ml glacial acetic acid and 15 ml 6 N HCl, respectively, in separate glasses), and are suspended in a fume-free chamber (possibly in a desiccator with a drying agent). The dried strips are washed twice with acetone, for 1 hr. each; after extraction, they contain the residual peptide and are ready for the next degradation step. The combined acetone extracts are evaporated and PTH-amino acids are taken up in, e.g., acetone.

For the extraction of PTH-histidine, PTH-cysteic acid, and PTH-arginine (these compounds are insoluble in acetone), half of a strip is extracted with a water/acetone mixture (5 : 95 w/w). Although this also brings the degraded peptide into solution, it does not interfere with chromatography. PTH-lysine presents some difficulties since this compound is not always extracted with acetone.

d) *Peptide Degradation According to Wieland [439].* A peptide mixture (1-5 mg) is applied in bands on a Silica Gel G plate and is developed with a suitable solvent (p. 98). Then the edges of the plate are sprayed with the color reagent. The peptide zone is now introduced into the reaction vessel (heavy-wall centrifuge tube with flat bottom, 4 cm diameter and 9 cm length), the gel is suspended with a small amount of dry acetone, is centrifuged and decanted, resulting in a uniform gel film. For the removal of traces of acetone, the tube is inverted and shaken, and the gel is subsequently impregnated with a 10% solution of PITC in 0.1 ml peroxide-free dioxane. The tube is now closed with a rubber stopper (protect with polyethylene) through which a glass rod is pushed. A small tube containing 5 ml of a solution of pyridine/dioxane/water (1 : 1 : 1) is attached at the bottom of the glass rod. The small tube should be at a distance of about 0.9 cm from the bottom. Heating to 40°C is carried out for 3 hrs. In order to remove the excess of PITC, the solution is extracted four times with 10 ml benzene (centrifuging and decanting of benzene). The open tube is held for 15 hrs. at 20°C in a desiccator containing glacial acetic acid and hydrochloric acid of constant boiling point in two separate dishes. The PTH-amino acids are extracted four times with 10 ml anhydrous acetone (centrifuging and decanting each time), and the combined extracts are evaporated to dryness in vacuum (compare pp. 153 ff). The dry gel film contains the degraded peptide and is ready for the second degradation step.

e) *Radioisotope Method of Cherbuliez [63, 67].* A micro-test tube (1.5-2 mm diameter and 40-45 mm length) is charged with 1-2 μl of S^{35}-PITC-solution in pentane (5 nM/μl; specific activity 30-80 mC/mM), is evaporated to dryness at room temperature, is immediately treated with 2 μl dimethylformamide, followed by 2 μl peptide solution (1-2 nM peptide in 0.05 M Veronal Sodium); the tube is closed, shaken, and incubated for 2 hrs. at 40°C. After addition of 2 μl 0.05 M Veronal Sodium solution, the excess reagent is extracted three times with 5 μl benzene/pyridine (9 : 1) and the aqueous phase is evaporated at room temperature (1-2 Torr). The residue is taken up in 5 μl glacial acetic acid/concentrated HCl (4 : 1) and the solution is allowed to react in the closed tube at 40°C for 2 hrs. Subsequently, it is repeatedly evaporated to dryness with several water additions (2 × 2 μl; 1-2 Torr at room temperature), the residue is dissolved in 4 μl 0.05 M Veronal Sodium/0.05 N NaOH (1 : 1), and the PTH-amino acids are extracted four times with 5 μl benzene. PTH-aspartic acid is isolated most suitably by extraction with 4 × 5 μl methylethylketone/benzene (3 : 2). The aqueous solution is evaporated to dryness before the next degradation step.

2. Degradation of Proteins

Excellent methods have been developed by Fraenkel-Conrat [119] (modification on p. 154), Eriksson and Sjöquist [107], and Laver [217] for the determination of the N-terminal amino acids in proteins.

von Korff et al. made a particularly detailed study of the quantitative determination [207].

a) *Protein Degradation According to Eriksson and Sjöquist [107].* A protein (5-20 mg) + 1 ml salt solution + 2 ml pyridine/triethyl-amine/PITC (100 : 30 : 1) are incubated for 90 min. at 40°C. Extraction is carried out five times with 5 ml benzene/ethylene chloride saturated with 0.1 N NaOH, followed by centrifuging, discarding of the organic phase, and evaporation of solvent rests under nitrogen at 40°C. For the cleavage of the N-terminal amino acid, the PTC-protein is treated with 1 ml water and 2 ml glacial acetic acid/saturated with HCl-gas for 2 hrs. at 40°C and is lyophilized. The residue is suspended in 2 ml ethylacetate/methylethylketone (2 : 1) saturated with water, the PTH-amino acids are extracted with the mentioned solution, the combined organic extracts are evaporated to dryness in vacuum under nitrogen and the residue is taken up in 0.1 ml 90% acetic acid. The accuracy of a quantitative determination (spectrophotometry of the eluted PTH-derivatives at 269 mμ and of the PTH-serine and PTH-threonine at 320 mμ) is increased if C^{14}-amino acids of known activity are used as internal standard [207] (see also *b*).

b) *Radioisotope Method According to Laver [217].* Protein (1-2 mg) are dissolved in 0.1 ml water, 0.1 ml buffer (1 g N-allylpiperidine + 39 ml pyridine, pH adjusted to 9 with 1 N acetic acid) is added, the mixture is treated with 5 μl S^{35}-PITC (specific activity 15 mC/mM), and shaken for 1 hr. at 40°C. The PTC-protein is extracted three times with 3 ml benzene, three times with 3 ml ethanol, and three times with 3 ml acetone (discard organic phases); the residue is dissolved in 0.2 ml formic acid and 0.2 ml 6 N HCl is added. After 5 min. at 20°C, 3 ml acetone are added, the protein is centrifuged off, the acetone is evaporated, and the residue is extracted with 2 ml acetone. The product is evaporated to dryness, the residue is dissolved in a drop of acetone, 0.2 ml 0.1 N HCl is added, and the solution is allowed to stand at room temperature for 15 hrs. It is evaporated and treated as described in (a).

C. CHROMATOGRAPHY OF PTH-AMINO ACIDS

1. Solvents and Separation Effects

Silica gel layers and solvents numbers 94 through 106 [44, 63, 64, 65, 114, 295], as well as aluminum oxide layers and solvents numbers 107, 108, and 109 [65], are suited for the chromatography of PTH-amino acids (Table 43).

Table 43. Solvent for the chromatography of PTH-amino acids.

No.	Solvent		Author
94[1]	Chloroform[6]		Brenner et al. [44]
95[1]	Chloroform[6]/methanol	9 : 1	
96[1]	Chloroform[6]/formic acid	100 : 5	
97[1]	Chloroform[6]/methanol/formic acid	70 : 30 : 2	
98[1]	"Heptane" [7]		Pataki [295]
99[2]	Chloroform[6]/petroleum ether/ glacial acetic acid	100 : 4 : 16	Fittkau et al. [114]
100[2]	Chloroform[6]/petroleum ether/ methanol	100 : 7 : 40	
101[2]	Chloroform[6]/methanol	10 : 4	
102[4]	Heptane/pyridine/glacial acetic acid	5 : 3 : 2	Cherbuliez et al. [63, 65]
103[4]	Chloroform[8]/isopropanol/water ..	28 : 8 : 1	
104[4]	Ethylacetate/pyridine/water	7 : 2 : 1	
105[4]	Chloroform[8]/ethylacetate/water ..	6 : 3 : 1	
106[5]	Chloroform[8]/ethylacetate	95 : 5	
107[3]	Chloroform[8]		
108[3]	Chloroform[8]/ethanol	98 : 2	
109[3]	Chloroform[8]/isopropanol/formic acid	35 : 4 : 1	

[1] Silica Gel G, air-dried; see text p. 65.

[2] Silica Gel (Agfa/Woelfen), see text, p. 66.

[3] 10 g aluminum oxide, Fluka (2 hrs. drying at 150°C) + 80 ml isopropanol (dried over Na_2SO_4) are vigorously shaken in an Erlenmeyer flask, 5.5 ml of the suspension are spread on a 5 × 20 cm plate, followed by 1-2 hrs. drying and possibly by activation.

[4] Silica Gel G; with solvents numbers 102 and 103, dry 2 hrs. at 130-140°C and with numbers 104 and 105, dry 2 hrs. at 40°C before chromatography.

[5] Small portions of 85 g silica gel (Woelm) are placed into a mortar filled with 15 ml water; 10 g of the deactivated adsorbent are vigorously shaken in an Erlenmeyer flask with 30 ml ethylacetate; 5 ml of this suspension are spread on a 5 × 20 cm plate. After 20-30 min. of drying in air, the plates are used immediately.

[6] Stabilized with 1.5% alcohol.

[7] n-heptane/ethylene chloride/formic acid/propionic acid (90 : 30 : 21 : 18); 100 ml of the upper phase are used.

[8] Pharmacopoea Helvetica V.

Brenner et al. [44] and Pataki [295] investigated the chromatographic behavior of 32 PTH-amino acids on Silica Gel G layers. Table 44 contains the R_f-values in solvents numbers 94 through 98.

Table 44. 100xR_f-values[1] of PTH-amino acids on Silica Gel G layers (see text p. 65). From Brenner et al. [44] and Pataki [293].

| PTH[2] | 100xR_f in solvent number | | | | |
	94[3]	95[3]	96[3]	97[4]	98[5]
Abut	26	79	54	—	—
Acy	44	84	67	—	—
α-Aib	27	80	56	—	—
Ala	18	77	44	~100	11
Arg	00	01	00	24	00
Asp	00	02	16	70	00
Asp(NH$_2$) ...	00	34	09	~100	00
Cit	00	34	08	—	—
CySO$_3$H	00	00	00	—	—
Glu	01	05	18	75	00
Glu(NH$_2$) ...	00	40	11	~100	00
Gly	11	68	35	90	05
His	01	40	01	~100	02
Hypro	05	64	28	—	—
Ile	39	83	62	~100	37
Leu	39	84	63	~100	37
Lys	12	78	34	~100	03
Met	34	81	54	~100	13
Me-Glu[6]	23	82	50	—	—
MetO	00	54	15	~100	01
MetO$_2$	02	59	17	—	—
Me-Ser[6]	01	51	18	—	—
Nleu	40	83	62	—	—
Nval	34	81	57	—	—
Orn	07	72	30	—	—
Phe	30	81	54	~100	18
Pro	60	89	70	~100	21
Ser	01	43	10	—	—
Thr	01	58	17	~100	00
Try	14	71	41	~100	10
Tyr	03	59	22	~100	01
Val	33	81	58	~100	23
MPTU[7]	12	65	32	—	03
DPTU[8]	42	82	71	—	22

[1] Average values of 6 single determinations are listed; the mean of the standard deviation amounts to 0.015 (see text, p. 48).

[2] Abbreviations of the amino acids according to Table 14.

[3] Development distance of the solvent: 18 cm.

[4] Development distance of the solvent: 11 cm.

[5] Development distance of the solvent: 15 cm.

[6] Methylglutamic acid and methylserine, respectively.

[7] Monophenylthiourea.

[8] Diphenylthiourea.

Cherbuliez et al. [63, 65] used Silica Gel G layers and solvents numbers 102 through 106 on one hand, and aluminum oxide layers and solvents numbers 107, 108, and 109 on the other. The R_f-values are shown in Table 45.

Recently Fittkau et al. [114] reported on the thin-layer chromatography of PTH-amino acids on silica gel (Agfa/Woelfen) layers (p. 66) with the use of solvents numbers 94, 99-101; however, they gave no detailed data. As shown by Table 45, quite useful separations can be

Table 45. R_f-values of PTH-amino acids on silica gel and aluminum oxide layers (compare Table 43). From Cherbuliez et al. [63, 65].

	$100xR_f$ in solvent number						
PTH[1]	103[2]	104[2]	105[2]	106[3]	107[4]	108[4]	109[4]
Ala	41	—	45	46	51	66	—
Arg	—	—	—	—	00	00	00
Asp	00	17	00	—	00	00	20
Asp(NH$_2$) ..	03	67	02	—	00	00	32
(CyS)$_2$	—	—	—	—	—	—	69
Glu	01	25	00	—	00	00	32
Glu(NH$_2$) ..	02	65	02	—	00	00	36
Gly	30	—	30	34	28	45	78
His	01	58	00	—	00	00	04
Ile	64	—	66	71	78	80	—
Leu	69	—	70	72	78	80	—
Lys	28	—	25	23	28	62	—
Met	52	—	51	57	58	76	—
Phe	56	—	53	60	60	74	—
Pro	75	—	61	81	91	91	—
Ser	06	79	07	05	00	00	48
Thr	11	83	14	09	00	00	51
Try	—	—	—	—	28	52	—
Tyr	19	—	35	16	00	08	64
Val	59	—	61	61	72	76	—

[1] Abbreviations of the amino acids according to Table 14.

[2] Silica Gel G (see Table 43).

[3] Woelm silica gel (see Table 43).

[4] Aluminum oxide, Fluka (see Table 43).

obtained even with a one-dimensional method. When three chromatograms are run simultaneously (Silica Gel G, solvents numbers 103, 104 and 105), it is possible to distinguish 15 of 17 PTH-amino acids (PTH-asparagine and PTH-glutamine remain unseparated); of 20 phenylthiohydantoins, 12 are separated in at least one of the three chromatograms with a simultaneous use of solvents numbers 107, 108, and 109 (aluminum oxide layer) (the PTH-derivatives of asparagine + glutamic acid, phenylalanine + methionine, leucine + isoleucine move together; the

development distances of PTH-serine and PTH-threonine differ little).
According to Brenner et al. [44, 295], the PTH-amino acids are
separated as follows. Two two-dimensional and one one-dimensional
chromatogram are prepared simultaneously on Silica Gel G layers (Fig.
78). PTH-glycine and monophenylthiourea (MPTU), as well as PTH-
leucine and PTH-isoleucine, remain unseparated; PTH-serine (not shown
in the figure) is found closely above glutamine, and PTH-methionine-
sulfone (also not shown) migrates together with threonine. PTH-glycine
can be detected beside MPTU with the aid of a specific color reaction
(p. 163); PTH-leucine and PTH-isoleucine can be distinguished on
Silica Gel G layers according to Cherbuliez [63] (solvent number 103 or
105; Table 45). Another possibility: chromatography (4x) on silica with
$CHCl_3/CH_3OH$ (9:1) [486].

2. Determination of the N-Terminal Amino Acid and Degradation of Peptides

Cherbuliez et al. [64] used thin-layer chromatography for the detection
of N-terminal amino acids in some di- and tripeptides and found a good
agreement between the R_f-value of the synthetic PTH-amino acid and
the R_f-value of the cleaved PTH-amino acid (Table 46). Thin-layer
chromatography permits the end-group determination of a di- or tripeptide with a quantity of about 0.01 μM sample.

Wieland and Gebert [439] used thin-layer chromatography in order
to investigate the sequence of desthiosecophalloidin A. Phalloidin was
desulfurated with Raney nickel and the desthiophalloidin was allowed
to stand for 2 hrs. in 50% aqueous trifluoroacetic acid. Desthiosecophalloidin A (1 mg) was dissolved in 0.1 ml methanol and the solution was
applied in bands on a Silica Gel G plate. The chromatogram was developed with methanol (development distance 15 cm), the band was

Table 46. Edman degradation of peptides: identification of the N-terminal
amino acid by chromatography on Silica Gel G layers. (Solvent number 102,
see Table 43; abbreviations according to Table 14; for the abbreviation of
peptides, compare [189].) From Cherbuliez et al. [64].

| Peptide | Quantity μM | μg | PTH-derivative of the N-terminal amino acid | | | | | |
| | | | Step 1 | | Step 2 | | Step 3 | |
			$R_f{}^1$	$R_f{}^2$	$R_f{}^1$	$R_f{}^2$	$R_f{}^1$	$R_f{}^2$
H-Gly-Pro-OH	0.02	3.44	0.370	0.365	0.515	0.515[3]	—	—
	0.01	1.72	0.375	0.375	0.495	0.495	—	—
H-Pro-Leu-Gly-OH .	0.02	6.42	0.490	0.490	0.615	0.615	0.385	0.390
	0.01	3.21	0.480	0.480	0.580	0.585	0.465	0.475

[1] Reference substance.
[2] PTH-derivative of the cleaved amino acid.
[3] + traces of PTH-glycine.

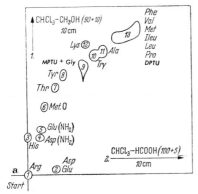

a) Note that only three of a total of 13 spots contain several components. For the separation of spots 2, 9, and 13, compare (b) and (c).

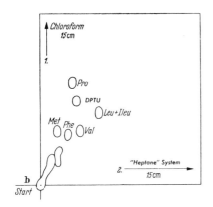

b) Note that the components of spot number 13 are separated with the exception of PTH-Leu and PTH-Ile. For the separation of these two compounds, see the text.

c) Separation of PTH-Asp and PTH-Glu (development distance 11 cm). In the case of complex mixtures, it is of advantage to elute spot number 2 of chromatograms according to Fig. 78a and to repeat the chromatography as described.

Fig. 78. Two- and one-dimensional chromatograms for the separation of PTH-amino acids on Silica Gel G layers. Load: 0.5 μg each in 0.5 μl methanol or acetone. Detection: chlorine/tolidine reaction or UV-light (see text, pp. 163 ff). From Brenner et al. [44] and Pataki [295].

scraped off at a distance of 4 cm from the start line and degraded five consecutive times as described on p. 155.

The PTH-derivatives of alanine, threonine, alanine, allohydroxyproline, and alanine could be identified by thin-layer chromatography practically without any impurities. Wieland and Gebert [439] used Silica Gel G layers and solvents numbers 94, 95, and 96 for this purpose.

Sarges and Witkop [359] used the degradation method of Eriksson and Sjöquist [107] in a somewhat modified form for the stepwise degradation of desformylgramicidin A. After the ninth degradation step, it was still possible to obtain an unequivocal detection of PTH-tryptophan.

The reaction was carried out in pyridine/triethylamine/PITC (100 : 3 : 1, a modification of p. 156) for 4 hrs. at 40°C. The excess of PITC was removed in a high vacuum and the residue was treated with trifluoroacetic acid at room temperature for 1 hr.; the PTH-amino acid was separated from residual peptide by column chromatography (Dowex 50 × 2, methanol, and 1 M ammonia as the eluant). The cleaved PTH-amino acid was determined each time by thin-layer and gas chromatography and the residual peptide was subjected to quantitative analysis. The technique given in Figs. 78a, b, and c has been reported by Habermann [465] to be superior to all other chromatographic methods for the determination of N-terminal amino acids.

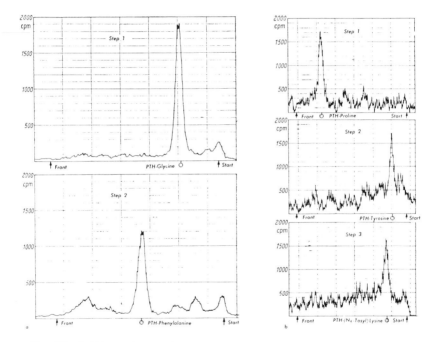

Figs. 79a, b. Edman degradation of peptides (quantity of substance, 1 nM each): activity distribution on thin-layer chromatograms (see text). The position of the PTH-amino acids was also determined by chemical reaction and is indicated as a shaded circle.

a) Degradation of H-Gly-Phe-OH.
b) Degradation of H-Pro-Tyr(Nε-tosyl)-Lys-OH. From Cherbuliez et al. [63].

For the detection of the degradation products of peptides which formed after S[35]-PITC treatment Cherbuliez et al. [63] used Silica Gel G layers and solvent number 103. Figs. 79a and b show the activity distribution during degradation of two peptides. Approximately 0.1 nM of a peptide are required for analysis with this method.

3. Detection Methods

a) *Chlorine/Tolidine Reaction.* PTH-amino acids can be detected by means of the chlorine/tolidine reaction (modification of Brenner et al. [44], p. 107; chlorine treatment is to be extended to 25 min.). The limit of detection is approximately $3 \times 10^{-4} \mu M$ [44].

b) *Butylhypochlorite/KI/Starch Reaction.* The air-dried plate is sprayed with a solution of 1% starch in water/0.1 N KI (1 : 1) and with a 1% butylhypochlorite solution in cyclohexane [114, 249].

c) *Iodine Azide Reaction.* The layer is sprayed with a 1 : 1 : 6 mixture of solutions I, II, and III. The mixture is stable for several days when kept in the cold [65].

Solution I: 1.27 g iodine + 8.3 g KI + water to 100 ml.

Solution II: 3.2 g sodium azide + water to 100 ml.

Solution III: 1% starch solution in water.

d) *UV-Light.* The PTH-amino acids are chromatographed on Silica Gel G/zinc silicate layers (30 g Merck Silica Gel G + 0.6 g zinc silicate;* for further details see p. 65), and the plate is examined in short-wave UV-light (254 mμ). Phenylthiohydantoins appear as dark spots on greenish fluorescent background. The limit of detection is about 6×10^{-4} μM PTH-amino acid [295].

e) *Radioisotope Method.* The reference substances which are co-chromatographed are detected as described above and, subsequently, the activity distribution on the plate is determined by means of a thin-layer scanner (p. 38) or another suitable method (Fig. 76).

f) *Specific Detection of PTH-Glycine.* The layer is sprayed sparingly with distilled water; it is then held over a concentrated ammonia solution (for example, an open ammonia bottle) and a deep-red stable spot appears. The limit of detection is about 0.08 μg PTH-glycine [295]. Threonine gives only a weak spot.

---

* Mn-activated zinc silicate of the Leuchtstoffwerke GmbH, Heidelberg, Germany; different fluorescent adsorbents are also marketed by the Merck and Camag companies.

Chapter 10

Trinitrophenyl Amino Alcohols

Trinitrophenyl amino alcohols (TNP-amino alcohols) form when proteins, peptides, or their esters are reduced with $LiBH_4$ (Chibnall and Rees [68]) and when the amino alcohol, after liberation by hydrolysis, is converted with 2,4,6-trinitrobenzene sulfonic acid (TNBS) (Satake et al. [360]). This reaction sequence permits a determination of the C-terminal amino acid of a peptide or protein.

Enzymatic or chemical methods can be used for the cleavage of the C-terminal amino acid (for a summary, see Bailey [6], Smyth and Elliot [397]).

Carboxypeptidase cleaves aromatic, as well as aliphatic, amino acids. The degradation reaction rate depends upon the C-terminal amino acid on one hand and upon the next to the last amino acid on the other; proline and hydroxyproline, as well as D-amino acids, are not cleaved. The amino acids which were liberated by enzymatic methods can be chromatographed on paper or on thin-layers in their free or in their dinitrophenylated form. Jutisz et al. [227] gave a brief report on the application of thin-layer chromatography for the detection of C-terminal amino acids.

In addition to the $LiBH_4$-reduction, the hydrazonolysis according to Akabori et al. [3] should be mentioned as a chemical method. Recently Jutisz et al. [229] reported on a modified method; thin-layer chromatography was used among other procedures.

Ishii and Witkop [179] described the use of thin-layer chromatography for the detection of TNP-amino alcohols.

Approximately 50 μM $LiBH_4$ in 50 μl dimethoxyethane were added to a solution of 8 mg seco-gramicidin A (or gramicidin A) in 1 ml dimethoxyethane; the turbid solution was allowed to react overnight (17 hrs.) at room temperature, was adjusted to pH = 3 with 1 ml dilute acetic acid, evaporated to dryness, and dissolved in 5 ml butanol + 5 ml water. The aqueous phase was extracted three times with 3 ml butanol,

164

the combined extracts were evaporated to dryness, and the residue was hydrolized in 2 ml 6 N HCl and 0.4 ml glacial acetic àcid at 110°C in an evacuated ampule. The hydrolysate was evaporated and dissolved in 0.4 ml water; ¼ of this solution was trinitrophenylated with 5 mg TNBS in 1 ml 0.5 N aqueous $NaHCO_3$ solution at 40°C for 2 hrs.; the reaction mixture was extracted twice with 2 ml ether and the ether solution was chromatographed on Silica Gel G layers. As a control, TNP-amino alcohols prepared as described above were run at the same time.

Table 47 contains the R_f-values of some TNP-amino alcohols according to Witkop [444].

Table 47. 100xR_f-values of some TNP-amino alcohols on silica gel layers [444] (see p. 65).

Compound	$100xR_{f1}$[1]	$100xR_{f2}$[2]	$100xR_{f3}$[3]
TNP-Alaninol	19	37	45
TNP-Isoleucinol	31	48	60
TNP-Leucinol	30	46	59
TNP-Valinol	28	43	58
TNP-Tryptophanol	14	28	44

[1] Chloroform/ethanol 100 : 1.
[2] Toluene/pyridine/glacial acetic acid 80 : 10 : 1.
[3] Benzene/glacial acetic acid 3 : 1.

PART IV

Amino Acids and
Related Compounds
in Biological Material

Chapter 11

Chromatography and Electrochromatography of Free Amino Acids

Since the fundamental research of Dent [91], numerous methods have been developed for the separation of amino acids in body fluid and organ extracts (for a summary see Bigwood et al. [30] and Smith [396]). The amino acids can be chromatographed in the free form or as derivatives. If the first alternative is preferred, interfering impurities (e.g., salts, proteins, lipids, etc.) must be removed before chromatography. Consequently, many authors prefer to convert the free amino acids into derivatives before chromatography, since they can then be separated from the impurities.

In clinical work, the determination of amino acids in urine and blood is of considerable importance (Schreier [367, 368], Berger [23, 24], Opienska-Blauth [288], and Milne [253]). In most cases it is not necessary to make a quantitative determination of urine and blood amino acids [23, 288]; many investigators therefore prepare qualitative chromatograms. The quantitative differences are then stated as an increase or decrease of the α-amino nitrogen. For exact quantitative determinations the double-labeling method of Beale and Whitehead (p. 61) is particularly suitable.

A. PREPARATIONS FOR CHROMATOGRAPHY

Body fluids and organ extracts usually contain peptides, protein, carbohydrates, urea, salts and lipoids in addition to the free amino acids. While the removal of lower peptides which are present in small quantities is usually impossible (compare p. 113), the remaining components can be separated more or less easily from the amino acids.

169

1. Removal of Proteins, Polysaccharides, and Lipoids

Different precipitation methods can be used to separate the high molecular weight components. Oepen and Oepen [282] have compared many *deproteination* methods and found that the type of deproteination

Table 48. Losses of the amino acid concentration after the addition of the test mixture and deproteinization with picric acid, 20% trichloroacetic acid and acetone. Quantities in μmole/ml serum. From Oepen and Oepen [282].

Deproteinization with	Picric acid	20% trichloro-acetic acid	Acetone
Taurine	0.006	0.009	0.026
Glutamic acid	0.045	0.012	0.013
Proline	0.015	0.028	0.044
Glycine	0.040	0.005	0.038
Alanine	0.046	0.006	0.033
Valine	0.015	0.001	0.008
Methionine	0.002	0.002	0.022
Isoleucine	0.019	0.017	0.025
Leucine	0.024	0.019	0.007
Tyrosine	0.009	0.036	0.012
Phenylalanine	0.006	0.035	0.000
Ornithine	0.006	0.010	0.029
Lysine	0.006	0.002	0.026
Histidine	0.003	0.002	0.030
Tryptophan	0.018	0.006	0.032
Arginine	0.010	0.013	0.025
Mean value	0.017	0.013	0.023

influences the amino acid composition. Table 48 contains the amino acid losses resulting from three different methods. The loss ratios of the added test mixtures were highest after treatment with acetone and lowest after treatment with trichloroacetic acid (Table 48).

According to Kraut and Zimmerman-Telschow [212], blood is centrifuged, the serum is diluted with water, and the high molecular weight components are precipitated with 20% trichloroacetic acid (serum/water/trichloroacetic acid = 1 : 1 : 1). After centrifuging and washing of the precipitate, the combined solutions are evaporated almost to dryness, are diluted with 0.1 N HCl, extracted with ether to remove lipoids and the ether is washed three times more with 0.1 N HCl. The combined hydrochloric acid solutions are evaporated to dryness in vacuum and the residue is taken up in 2 ml 0.1 N HCl. The high molecular weight compounds can also be simply separated by means of Sephadex gel filtration (p. 113). This method causes little alteration of the material investigated. Berger et al. [25] were unable to detect cysteine/cystine after deproteination with Sephadex. They attribute this fact to the mild deproteination with Sephadex and are of the opinion that the cystine (and possibly also cysteine) found by other authors in the serum is cleaved for the most part from the serum proteins during acid deproteination.

2. Decomposition of Urea

By the addition of a trace of urease, it is possible to convert urine-urea into CO_2 and ammonia within 24 hrs. These substances escape with steam during a subsequent concentration [34].

3. Desalting

Body fluids and organ extracts frequently contain considerable quantities of salt. If the salt concentration is high compared to the amino acids, it is necessary to remove them before chromatography (p. 169). In electrolytic desalting [73], arginine is partially converted into ornithine; histidine, lysine, methionine, proline and tyrosine are lost in amounts of 10-30%. Comparative studies of Opienska-Blauth [131, 284] showed that different cationic exchange resins are suited for desalting and furnish comparable results provided that suitable eluants are selected. Taurine is obtained with the combined use of cationic and anionic exchangers.

Bujard [55] recommends the following method for desalting of milk. Milk is concentrated somewhat and deproteinized with 20% trichloroacetic acid (milk TCE = 1 : 1) (p. 170). The supernatant solution is decanted and shaken for 1 hr. with Amberlite 120 × 8 (15-30 mesh). The resin is washed repeatedly with water (decanting) and the excess of water is carefully aspirated with a pipette. In order to obtain the salt-free amino acids, the resin is treated with a small quantity of 4 N ammonia and shaken for 25 min.; the ammonia is carefully suctioned off and elution is completed with 0.5 ml 4 N ammonia. The combined ammonia solutions are evaporated on a watch glass over a water bath and the dry residue is taken up in water.

The simultaneous removal of salts, proteins and carbohydrates is possible according to the following method of Harris et al. [153]: Dowex 50 W (× 8, 200-400 mesh, analytical grade) is shaken for 2 min. with a fivefold quantity of water, the supernatant suspension is suctioned off, and this treatment is repeated twice. Before use the resin is suspended with 1 N HCl, water, 1 N HCl, and is finally washed free from chloride with water. The product is now dried on filter paper (40% water); in this state, 1 g resin contains about 3 meq. sulfonic acid groups. For 800 μeq. inorganic cations + 20-50 (but a maximum of 100) μM amino acid, about 1050 μeq. resin are required (resin excess of 200 μeq.). A column of 5 mm inside diameter is filled with 350 mg wet resin (to be converted into the H-form with 5 eq. 1 N HCl; the capacity is determined by allowing excess NaCl solution to flow through and titrating the hydrochloric acid formed). Serum or plasma is diluted with an equal volume of water, is adjusted to pH = 2-2.5 with 8 N acetic acid (remove CO_2), and is charged on the column. After the sample solution has passed through, the column is washed with 3 ml 0.5 N acetic acid and 2 ml water. The amino acids are eluted with 2 N triethylamine solution (triethylamine in 20% acetone is stable for 2 weeks in the cold in a well-closed bottle).

The eluate is tested by the addition of traces of a Bromcresol green solution to each drop until the pH-value just begins to rise ("breakthrough volume" = x). If x is not greater than 0.5 ml, it does not need to be discarded. Usually a quantity of solvent of x + 1 ml is sufficient. After breakthrough, the fractions are collected for analysis. Practically all amino acids are contained in the first 1 ml fraction. The solution is concentrated in siliconized porcelain crucibles over sulfuric acid at room temperature and is dissolved in 90-100 μl water. The recovery of amino acids is approximately quantitative. A limitation of this elegant method is that taurine and cysteic acid are eluted together with the wash liquid (see above). If necessary, the basic amino acids + tryptophan can be separated from the remaining amino acids [153].

4. Determination of α-Amino Nitrogen

Qualitative chromatography gives a picture of the presence of amino acids in body fluids. The degree of aminoaciduria or aminoacidemia, for example, is determined by calculating the α-amino-N and the amino-N-index.

$$\left(\frac{\alpha\text{-NH}_2\text{-N}}{\text{total N}} \times 100 \right)$$

(Berger [23, 24], Opienska-Blauth [284]). According to Berger, the somewhat modified method of Antener [4] is suited for the determination of α-amino-N.

Determination of α-amino-N in blood or serum [24]: 6 ml tungstic acid (40 ml 10% sodium tungstate solution are diluted with 700-800 ml distilled water, 40 ml ⅔ N sulfuric acid are added, and the volume is brought to 1 l with water; the solution must be replaced as soon as a white precipitate forms) are placed into a centrifuge tube (20 ml). 0.1 ml blood is added; the solution is mixed and shaken, allowed to stand for 5 min., centrifuged for 5 min. at 3600 rpm, and filtered through a small cotton ball. Of the filtrate, 5 ml are introduced into a glass tube (8 cm high and 1.5 cm diameter) which has a mark at 10 ml; 5 ml distilled water are used for the blank value. Now 0.05 ml phenolphtalein solution (0.25% in ethanol) is added, followed by 0.05 N NaOH from a microburette until a distinct red color indicates the endpoint. Now 1 ml

2% Borax solution and 0.4 ml β-naphthoquinone solution (0.05% sodium β-naphthoquinonate freshly prepared each time) are added and the mixture is placed into boiling water for precisely 3 min. The solution is cooled under flowing water, is well-mixed with 1 ml acetic acid/sodium acetate solution (100 ml 50% acetic acid + 100 ml 5% sodium acetate) and with 1 ml 4% sodium thiosulfate solution, is brought to 10 ml with distilled water, and is subjected to colorimetry (e.g., Pulfrich Photometer, filter S 50, 1 cm layer thickness). The standard solution contains 0.1 mg amino-nitrogen per ml (187.8 mg proline + 412.8 mg glycine dissolved in 1 l of 0.07 N HCl containing 0.2% sodium benzoate). It is advisable to perform double-determinations and to record a calibration curve for each series of experiments. The boiling time (3 min.) must be exactly observed; the addition of acetic acid and thiosulfate, furthermore, must be made immediately after cooling and readings are taken directly afterwards (not later than after 15 min.).

Determination of α-amino-N in urine [24]: 24-hr. urine is stabilized with thymol and, if necessary, is stored at 4°C. It is filtered through folded filter paper (protein-containing urine is treated with the same volume of 10% sulfosalicylic acid), 10-20 ml are removed, the pH is adjusted to 1 with 10% HCl (indicator paper)—concentrated NaOH is used to adjust the pH of de-proteinized urine—the uric acid is removed by centrifuging, and the residue is filtered (uric acid precipitates if the mixture is allowed to cool for about 25 min. in an ice/salt mixture). Of the filtrate, 1 ml is charged into a special heavy-walled glass attachment, 1 drop concentrated NaOH is added and the mixture is evacuated in vacuum with gentle heating (about 40°C) of the liquid layer (careful —foaming). The nearly dry residue is quantitatively flushed with distilled water in a 100 ml graduated cylinder (50 ml for a 1 : 1 dilution of urine) and is brought to a volume of 100 ml. A quantity of 5 ml removed, adjusted to phenophthalein red, and the procedure continues as with serum.

Some drugs (e.g., sulfonamides, p-aminosalicylic acid) interfere with the analysis. Sugar, acetone, urea, creatine, and bilirubin, on the other hand, do not interfere.

The Ninhydrin method modified by Müting and Kaiser [269] is also suited for a routine determination of α-amino-N. The ammonia, urea, and uric acid are of little influence.

B. SEPARATION OF AMINO ACIDS

1. Amino Acids in Urine and Blood

Opienska-Blauth et al. [289], Baron and Economidis [10], Rokkones [346], as well as Crawhall et al. [74], have investigated the thin-layer chromatographic detection of urine amino acids.

If not more than 10 μl urine are applied, desalting is not absolutely necessary [10, 289, 346]. The best thin-layer chromatograms are obtained when about 10 μl native urine or about 50 μl demineralized urine are applied (Opienska-Blauth [289]). Opienska-Blauth et al. [289] chromatographed urine amino acids on Silica Gel G layers with solvents numbers 8 and 9 using the two-dimensional technique (p. 71 and Table 12) or on Whatman number 3 paper with the same solvent combination. If about 1-10 μl of a physiological urine are applied, followed by Ninhydrin treatment, no spots are visible on the paper while 3 to 6 can be seen on the layer (Fig. 80a); with a load of 100 μl demineralized urine, 9-10 spots can be observed on paper chromatograms and 14-16 on thin-layer chromatograms. Baron and Economidis [10] used the same solvent combination and were able to detect 8 amino acids in 10 μl of a normal urine without desalting (Fig. 80b). Rokkones [346] prefers the combination of solvents numbers 13 and 9 (p. 79 and Table 12) and uses Silica Gel G layers to apply a quantity of urine corresponding to 20 μg creatinine (with or without demineralization; the former procedure results in better chromatograms). In healthy children of 2-3 years of age, 9-12 Ninhydrin-positive compounds are visualized (Fig. 80c). Fig. 80d shows the amino acids in 25 μl demineralized urine according to Opienska-Blauth [289]. Recently, Ambert et al. [458] have reported in great detail on the separation of urine amino acids.

The results of Rokkones [346] on one hand and of Opienska-Blauth [289] on the other can be compared only with reservations, since difference desalting methods were used (electrolytic desalting and Amberlite IR-120). In patients with cystinuria (elevated cystine and lysine urine levels), Crawhall et al. [74] analyzed cystine quantitatively on one hand and chromatographed the urine amino acids on Silica Gel G layers on the other (chloroform/methanol/17% ammonia, 2 : 2 : 1; and phenol/15% aqueous formic acid solution, 75 : 25). The mentioned solvent combination permits a good separation of ornithine, lysine, and arginine from each other and also separates these compounds from the other amino acids (p. 74). An electrochromatographic method for the separation of urine amino acids has recently been described [482].

Rokkones [347] gives the following procedures for a semi-quantitative analysis of urine amino acids: 20 mixtures are prepared of 18 amino acids (composition and concentrations are shown in Table 49); 5 μl of each mixture are applied on Silica Gel G layers and development is performed with solvent number 9 with the one-dimensional technique (Table 12); the plate then contains 4 spots of different concentration and intensity for each amino acid. Such a chromatogram is carried out together with a test series. The urine amino acids are subjected to two-dimensional chromatography (Silica Gel G, solvents numbers 13 and

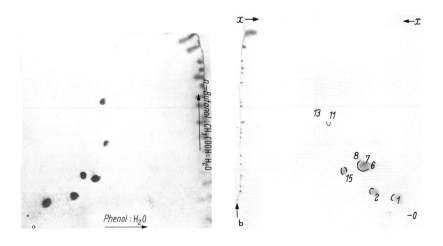

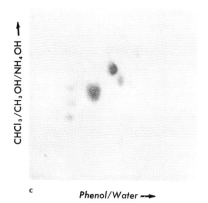

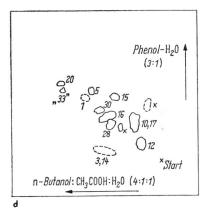

Fig. 80. Thin-layer chromatography detection of amino acids in urine. First dimension: n-butanol/glacial acetic acid/water 4 : 1 : 1 (a, b, and d) and chloroform/methanol/17% ammonia 2 : 2 : 1 (c) respectively; second dimension: phenol/water 75 : 25 w/w (compare pp. 71, 74); layer: Silica Gel G.

a) Load: 10 μl native urine. From Opienska-Blauth [289].

b) Load: 10 μl native urine; 1 = Lys, 2 = His, 6 = Ser, 7 = Gly, 8 = Ala, 11 = Tyr, 13 = Phe, 15 = Glu(NH₂). From Baron and Economidis [10].

c) Demineralized urine corresponding to 20 μg creatinine. From Rokkones [346].

d) Load: 25 μl deionized urine; 3 + 14 = Asp + Glu, 12 = (Cys)₂, 10 + 17 = Arg + His, 28 = Ser, 16 = Gly, 30 = Thr, 15 = Glu(NH₂), 1 = Ala, 5 = Abu (?), "33" = Val (compare p. 72), 20 = Leu + Ile. From Opienska-Blauth [289].

9, p. 79 and Table 12) and the spot intensities are compared. For comparative studies, i.e., for qualitative or semi-quantitative or quantitative analyses, identical aliquots of the 24 hr. urine—or those quantities of urine which correspond to a certain concentration of α-amino-N or

Table 49. Test mixtures[1] for the semiquantitative determination of urine amino acids [347]; 5 μl of each mixture are applied.

	Mixture I_1 mg/100 ml	Mixture I_2 mg/100 ml	Mixture I_3 mg/100 ml	Mixture I_4 mg/100 ml
Phosphoethanolamine .	20	50	100	200
Glutamine	20	50	100	200
Phenylalanine	4	10	20	50
Leucine	4	10	20	50
	Mixture II_1	Mixture II_2	Mixture II_3	Mixture II_4
Hydroxyproline	10	20	50	100
Cysteine	4	10	20	50
Taurine	20	50	100	200
Tryptophan	2	4	10	20
	Mixture III_1	Mixture III_2	Mixture III_3	Mixture III_4
Serine	10	20	50	100
Alanine	10	20	50	100
Valine	20	50	100	200
	Mixture IV_1	Mixture IV_2	Mixture IV_3	Mixture IV_4
Lysine	10	20	50	100
Glycine	10	20	50	100
Histidine	10	20	50	100
3-methylhistidine	4	10	20	50
	Mixture V_1	Mixture V_2	Mixture V_3	Mixture V_4
Proline	10	20	50	100
β-aminoisobutyric acid	10	20	50	100
Tyrosine	4	10	20	50

[1] The amino acids are dissolved in 10% isopropanol.

creatinine—must be applied (determination of α-amino-N, see p. 172; creatinine, see pp. 177-178). Rokkones [347] used the semi-quantitative method to investigate the excretion of urine amino acids in different age groups (load: urine corresponding to 20 μg creatinine) and found that some Ninhydrin-positive substances show a relation to age. Glycine, cystine, histidine, alanine, lysine, serine, glutamine, phenylalanine, and tryptophan as well as a few unidentified compounds are often excreted in maximum amounts per g creatinine in the first months of life (less than 6 months). These amino acids reach an "adult value" in 2-4 years. The 3-methylhistidine and taurine excretion is independent of age with the exception of the first days following birth. In the first days of life, the eliminated quantities of proline and hydroxyproline are barely measureable; the maximum concentration per g of creatinine is attained before the first month and decreases in about 4-5 months to undetectable concentrations. A higher cystine excretion usually is proportional to the lysine excretion; higher phenylalanine, tyrosine, and tryptophan excretions generally are combined in a similar manner. A comparison of the amino acid concentration in the urine of younger and older adults showed no significant differences; only the histidine excretion seems to decrease somewhat in older patients; however, some unidentified compounds were detected, the excretion of which per g of creatinine exhibited a clear relation to age. The findings of Rokkones [347] might facilitate the interpretation of chromatographic results even by less well-trained investigators.

Little is yet known about a thin-layer chromatographic detection of amino acids in blood. Opienska-Blauth [285] detected 14 Ninhydrin-positive spots in 20 μl plasma (chromatography on MN-300-cellulose layers with solvents numbers 8 and 9, p. 71 and Table 12). Fig. 81a shows the spot pattern.

Von Euler et al. [108] chromatographed normal and tumor serum of rats after deproteination with methanol, using Silica Gel G layers and n-butanol/glacial acetic acid/water (3 : 1 : 1) as solvent; they found that a spot with $R_f = 0.23$ appears in considerably higher concentration in the tumor serum. A volume corresponding to 40 μl deproteinized serum as well as different quantities of the two sera and reference substances of deproteinized tumor serum were then applied on the plate and after developing the chromatogram (Fig. 81b), the questionable zone was eluted with 2 ml water (20 min. at 80°C).

After *rechromatography* in two different chromatographic systems, the eluted amino acid could be identified as glycine. Dimillier and Trout [96] investigated the variation of blood amino acids during extracorporeal recirculation of blood (one- and two-dimensional chromatography on Silica Gel G layers) and observed that the concentration of many amino acids—e.g., histidine, proline, hydroxyproline, serine, threonine, aspartic acid, glutamic acid, and tryptophan, increases as a function of the pumping time. A considerably smaller increase of concentration was found in lysine and arginine; no significant variation of leucine, isoleucine, and methionine could be observed. Since it is known that hyperaminoacidemia is toxic, this finding probably is of clinical, as well as of biochemical, significance.

It has already been pointed out that the amino acid excretion is often referred to the creatinine concentration. The usual Jaffe reaction used for the detection and quantitative determination of creatinine is not specific. For example, several Jaffe-positive substances are observed in urine in addition to creatinine (Pataki [304]) (Fig. 82). The specific determination of creatinine in urine and blood is of considerable diagnostic importance in renal function tests.

The following method is suited for a quantitative analysis of cre-

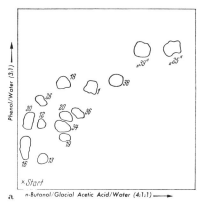

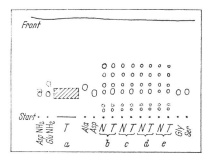

a) Layer: MN-300-cellulose; load: 20 μl plasma; 13 = Asp, 16 = (Cys)₂, 30 = Orn, 12 = Asp(NH₂), 19 = Glu, 34 = Ser, 20 = Gly, 26 = Lys, 36 = Thr, 18 = Glu(NH₂), 1 = Ala, 38 = Tyr, "39" = "Val" (see p. 85), "25" = Leu + Ile. From Pataki [305].

b) Solvent: n-butanol-glacial acetic acid/water 3 : 1 : 1; layer: Silica Gel G; a = 40 μl serum, b = 2 μl, c = 1.6 μl, d = 1.2 μl, e = 0.8 μl. The shaded area was not sprayed (see text). T = tumor serum, N = normal serum. From von Euler et al. [108].

Fig. 81. Thin-layer chromatography detection of amino acids in the serum and plasma [285, 108].

atinine in urine (Pataki and Keller [310]): 10 μl urine (A), 10 μl urine/distilled water 1 : 1 (B), and 10 μl creatinine standard containing 10 μg creatinine (S) are applied on a Cellulose-D plate (p. 76). After chromatography with n-butanol/water/glacial acetic acid (4 : 1 : 1) (Fig. 82), the plate is dried and is sprayed with 5% ethanolic picric acid solution and with 10% aqueous sodium hydroxide solution. After full color development, the creatinine spots are copied on vellum paper; the spot area and the creatinine concentration are determined according to the instructions on pp. 57-58. After elution, simultaneous quantitative determination of creatinine in urine and blood is done by photometry. For this purpose, about 10 μl urine or 300 μl serum are applied on the

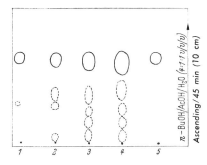

Fig. 82. Thin-layer chromatographic detection of creatinine in urine on cellulose-D layers (see p. 76); 1 = 5 μl, 2 = 10 μl, 3 = 15 μl, 4 = 20 μl urine, 5 = 5 μg creatinine. Note the secondary spots below creatinine; the intensity of these spots varies highly from one urine specimen to another: in some urine specimens they are entirely absent. From Pataki [304].

layer in spots or bands (serum is deproteinized with the same volume of acetone), and a standard solution is chromatographed at the same time; two standard substances at the plate edges are carefully colored, the creatinine zones are scraped off (analysis, standard, and an equal area as blank value), and the creatinine is eluted from the layer. Elution can be performed with 2 ml 0.1 N HCl (shaking for 3 hrs.), followed by neutralizing with 2 ml 0.1 N NaOH and addition of 2 ml color reagent [100 ml 1% (w/v) picric acid mixed with 40 ml 10% (w/v) sodium hydroxide solution and filled to 1000 ml with distilled water]. The color is stable for about 1 hr. The extinction is measured at 490 mμ (2 cm layer thickness).

2. Additional Applications

Thin-layer chromatography of amino acids can also be a valuable aid in a variety of pharmacological problems. Marquardt et al. [244] detected considerable quantities of histamine by thin-layer chromatographic methods in different varieties of wine and beer but not in grape juice. This finding explains the occasional complaints occurring after the consumption of wine (e.g. headache, shortness of breath, palpitations, and hyperacidity).

Carisano [60] used Silica Gel G layers for the detection of 3-methylhistidine, as well as larger quantities of histidine, for the distinction of whale meat and beef extracts. In addition to 15 μg histidine, 1 μg 3-methylhistidine can still be detected in two-dimensional chromatograms (first dimension: methanol/pyridine/water/glacial acetic acid, 6 : 6 : 4 : 1; second dimension: phenol/ethanol/water/ammonia, 3 : 1 : 1 : 0.1) (detection according to Moffat and Lytle, p. 92; histidine and 3-methylhistidine are given different colors). The free amino acids in milk can be detected on MN-300-cellulose layers in two-dimensional chromatograms (see Fig. 63) according to Bujard [55] (for the preparation for chromatography, see p. 65). Brieskorn and Glasz [52, 138] isolated hydroxylysine as a free amino acid and α-aminobutyric acid as a protein building block from the seeds of *Salvia officinalis L* .These two compounds could also be identified by thin-layer chromatography (Silica Gel G layers; tetrahydrofuran/pyridine/water, 3 : 1 : 1 as solvent; $R_{fHylys} = 0.06$, $R_{fAbut} = 0.65$). Huber et al. [173] determined the free amino acids and total amino acids in different corn steep water charges and found larger differences in some cases. Chromatography was carried out on Supergel layers (p. 65) with solvents numbers 13 and 9. Fig. 83 shows the spot pattern.

Michl and Bachmayer [251] used Silica Gel G layers for the two-dimensional separation of free amino acids in the toxin of the Bombina variegata L. and detected 13 amino acids and 5-hydroxytryptamine (first dimension: chloroform/methanol/17% ammonia, 2 : 2 : 1; second dimension: phenol/water, 3 : 1 w/v). The spot pattern has already been shown in Fig. 61c. A direct chromatography of the deproteinized toad toxin did not furnish reliable results since the low molecular weight peptides partly interfered with the amino acids. Consequently, the amino acids had to be concentrated [251].

Squibb [400, 401] reported on the quantitative determination of free amino acids in liver extracts; this densitometric method has already been discussed on pp. 59 and 71.

Nybom [281] separated the amino acids from the juices of *Fragaria* leaves by two-dimensional electrophoresis-chromatography on MN-300-

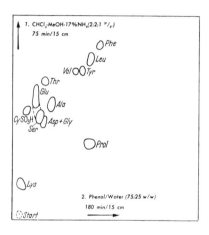

Fig. 83. Thin-layer chromatographic detection of amino acids in corn steeping water on Supergel layers (see p. 65). Note that the spot pattern partly deviates from the data of Fig. 61. From Huber et al. [173].

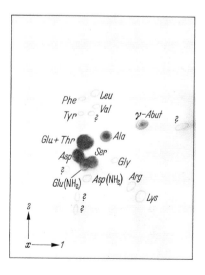

Fig. 84a. Electrophoresis-chromatography: amino acids from *Fragaria* leaves; layer: MN-300-cellulose; first dimension: electrophoresis (see Table 23, No. II, development time: 15 min.); second dimension: chromatography with n-butanol/glacial acetic acid/water 4 : 1 : 2 (2 hrs.). From Nybom [281].

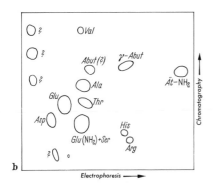

Fig. 84b. Electrophoresis-chromatography: Amino acids from sugar beet leaf juice [413]; layer: MN-300-cellulose; first dimension: electrophoresis (1200 V/20 cm, 25 min., pH 3.9; buffer = 240 ml glacial acetic acid + 60 ml pyridine + water to 5 l); second dimension: chromatography with n-butanol/glacial acetic acid/water 4 : 1 : 1 continuous flow development: 15 hrs.; ascending flow chromatography; compare p. 20).

cellulose layers. Electrophoresis was carried out in the first dimension (Table 23, No. II) followed by chromatography in the second dimension (butanol/glacial acetic acid/water, 4 : 1 : 2). The spot pattern is shown in Fig. 84a. Separation was successful without previous purification.

Stegemann and Lerch [414] separated amino acids from the leaf juices of sugar-beet by means of electrophoresis-chromatography (electrophoresis at pH = 3.9, 1200 V/20 cm, for 25 min. in the first dimension and continuous flow chromatography with butanol/glacial acetic acid/water, 4 : 1 : 1, for 15 hrs. in the second dimension). Fig. 84b shows the spot pattern. For additional examples, cf. [459, 483].

Chapter 12

Chromatography of Dinitrophenyl Amino Acids

Just as paper chromatography does, thin-layer chromatography also frequently fails in the investigation of biological specimens. The interference caused by the salt effect in these cases can be avoided if the amino acids are converted into derivatives which, because of their solubility in organic solvents, can be separated from the salts before chromatography based on their solubility in organic solvents. Suitable reagents are 2,4-dinitrofluorobenzene (p. 126), 2,4-dinitrofluoroaniline (p. 144), 1-dimethylaminonaphthalene-5-sulfonic acid (p. 144), and phenylisothiocyanate (p. 149). The conversion of free amino acids into dinitrophenyl (DNP) derivatives by means of 2,4-dinitrofluorobenzene (DNFB) can be rapidly carried out without tedious preparations even in the most highly dilute solutions. No data have yet been published on the application of the other cited reagents for the separation of amino acids from interfering components.

A. PREPARATION FOR CHROMATOGRAPHY

DNP-amino acids are sensitive to light (p. 126). Consequently, all manipulations should be carried out in the absence of direct light and if possible in darkness.

1. Dinitrophenylation

a) *Amino Acids in Urine [45, 434].*   A quantity of 25 ml fresh urine is treated with 5 N NaOH in drops up to a weak pink color of phenolphthalein paper followed by filtering through a dry folded filter, treatment of 20 ml filtrate with 5 ml buffer (8.4 g NaHCO$_3$ in water is filled to 100 ml with 5 N NaOH and distilled water; pH = 8.8) and with 2 ml freshly prepared, absolute-alcohol (10% w/v) DNFB-solution; 40 ml absolute

182

ethyl alcohol are added and the mixture is shaken or stirred in darkness for 1 hr. at 40°C. The reaction mixture is cooled to room temperature, adjusted to pH ~ 12, and transferred with a small quantity of distilled water as well as with about 10 ml ether into a 100 ml separatory funnel. The mixture is shaken thoroughly, the aqueous phase is separated and extracted twice with 10 ml ether each (discard ether extracts). If a gel-like, air-bubble–filled material makes the phase separation in ether difficult, the larger part of the aqueous phase is first separated, the remaining liquid is poured through glass wool into another separatory funnel, the wool is flushed with a little ether and weakly alkaline water (phenolphthalein-pink). The final phase separation now no longer presents difficulties; the aqueous fraction is combined with the main quantity of the aqueous phase. The same procedure may be necessary in the extraction of DNP-amino acids (2a) from an acidified aqueous solution (wash water adjusted to congo-acidity with hydrochloric acid) which is described below.

b) *Amino Acids in Blood [306, 311].* Serum (2-5 ml) is deproteinized with a twofold volume of acetone or with another suitable reagent (p. 170), adjusted to phenolphthalein-pink with 5 N NaOH, and treated with 5 ml carbonate buffer [see a)] and with 0.2 ml DNFB-solution [see a)] as well as with 15 ml absolute alcohol. The procedure continues according to the instructions in a).

c) *Amino Acids in Sperm [196, 306].* Ejaculate (2 ml) is shaken with 2 ml acetone and centrifuged for 15 min. The clear solution is adjusted to phenolphthalein-pink with 5 N NaOH; 3 ml carbonate buffer see a), 1 ml DNFB-solution [see a)] as well as 8 ml absolute alcohol are added and the procedure continues as in a).

2. Extraction of DNP-Amino Acids [196, 311, 434, 45]

a) *Ether-Soluble DNP-Amino Acids.* The alkaline-ether extracted reaction solution [1a), 1b), or 1c)] is carefully treated with 6 N HCl up to a clear blue color of congo paper, followed by extraction with 10 ml ether each six times and evaporation of the combined ether solutions in vacuum to dryness. The residue is dissolved in about 1 ml acetone or ethylacetate.

b) *Water-Soluble DNP-Amino Acids.* The acid solution remaining after the extraction of the ether-soluble DNP-compounds is extracted six times with 10 ml n-butanol/ethylacetate (1 : 1), the combined organic solutions are evaporated to dryness, and the residue is taken up most suitably in 1-2 ml ethylacetate/glacial acetic acid/n-butanol (100 : 1 : 99). After applying the solution on the layer, the acid must be completely evaporated before the start of chromatography.

B. SEPARATION OF DNP-AMINO ACIDS

1. Amino Acids in Urine, Blood, and Sperm

Brenner et al. [45, 434] investigated the thin-layer chromatography of urine amino acids. For more recent results, see also Figge [462]. The separation of the ether-soluble DNP derivatives from normal urine is shown in Fig. 85 (p. 137, method B).

In order to interpret chromatograms as shown in Fig. 85, different

Table 50. Free amino acids in urine, serum, and sperm [434, 311, 196]. (See text, p. 182.) +++ = larger quantities; ++ = medium quantities; + = small quantities; s = trace.

Amino acid	Urine[1]	Serum[2]	Sperm[3]
Alanine	+++	+++	++
Cystine	+++	−	−
Glutamine	+++	+++	+++
Glycine	+++	+++	+++
Histidine	+++	−	+++
Serine	+++	++	+++
Lysine	+++	+	s
β-aminoisobutyric acid ..	++	−	s[4]
Asparagine	++	−	−
Phenylalanine	++	+	++
Threonine	++	++	++
Tryptophan	++	−	s[4]
Tyrosine	++	+	++
Glutamic acid	+	++	++
Isoleucine	+	+	+++
Leucine	+	+	+++
Methionine sulfone	+	s	s
Ornithine	+	++	s[4]
Valine	+	+++	+++
α-aminoadipic acid	s	−	−
α-aminobutyric acid	s	s	−
β-alanine	s	−	s[4]
Aspartic acid	s	s	++
Proline	s	+++	+
Taurine	++	s	+
Arginine	++	s	+++
Cysteinic acid	+	s	+
Citrulline	s	−	−
γ-aminobutyric acid	−	−	+

[1] In 80 μl morning urine (ether-soluble derivatives) and 50 μl morning urine (water-soluble derivatives).

[2] 60 μl serum.

[3] 50 μl sperm.

[4] 100 μl sperm.

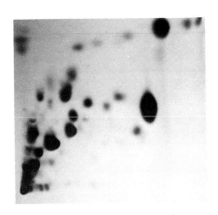

Fig. 85a (see p. 182). Load corresponding to 80 μl morning urine; "UV-photography"; experimental conditions according to Fig. 74a; for the interpretation of the spots, see Fig. 74a and Fig. 86 (see also Table 50). From Brenner et al. [434].

Fig. 85b (see p. 182). Load corresponding to 80 μl morning urine; "UV-photography"; experimental conditions according to Fig. 74b; note the separation of Di-DNP-Tyr and -Lys; for the interpretation of the spots see Fig. 74b (compare Table 50) [306].

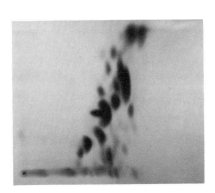

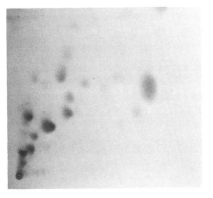

Fig. 85c. Load corresponding to 80 μl morning urine; "UV-photography"; separation of the DNP-derivatives of cystine, asparagine, glutamine and of aspartic acid, glutamic acid, and α-aminodipic acid according to Fig. 74c (compare Table 50) [306].

Fig. 85d (bottom). "Daylight photography": Compare Fig. 85a; in both cases, the same chromatogram is involved. From Brenner et al. [434].

Fig. 85a-d. Thin-layer chromatographic detection of the ether-soluble DNP-amino acids from urine (compare text) on Silica Gel G layers. For the photography technique, see p. 142.

spot patterns are prepared by chromatographing a standard mixture (for each case, 1 mg DNP-amino acid in a total of 5 ml ethylacetate or acetone; load: 2 μl = 0.4 μg per compound). For the assignment of spots, see Figs. 74a, b, and c. A comparison of Figs. 85 and 74 furnishes the amino acid composition of the investigated urine (Table 50). More recent investigations of Pataki and Keller [309] have shown that a total of 45 DNP-compounds can be detected in the ether-soluble fraction (Fig. 86 and Table 51). The spots designated by X could not be assigned

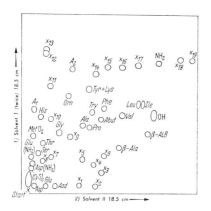

Fig. 86. General chromatogram of the ether-soluble DNP-compounds detected in urine (compare Table 51). Experimental conditions according to Fig. 74a; compare also Fig. 85a. A_1 and A_2 are artifacts. From Pataki and Keller [309].

to any of the investigated DNP-amino acids. In most cases, it is probable that peptides are involved. For if the urine is hydrolized with 6 N HCl before dinitrophenylation, a number of the unknown spots disappear in the chromatogram (Fig. 86); it must be pointed out, however, that new, partly unknown "spots" also form here. Recently, Bürgi et al. [57a] were able to detect numerous other unknown compounds.

The separation of water-soluble DNP-acids takes place by the one- or two-dimensional technique on Silica Gel G layers [434, 468]. Development is carried out three consecutive times with pure pyridine in the first dimension (the layer is air-dried for 3 min. after each run) and with n-butyl-alcohol saturated with 25% ammonia at room temperature in the second direction (after intermediate drying for 10 min. in an air draft; 10 min. at 60°C, and 15 min. in an air-draft). Fig. 87a shows the water-soluble DNP-amino acids of normal urine; assignment was made according to Fig. 87b (Table 50). Bürgi et al. [57a] described another method. The water-soluble DNP-compounds from urine are shown in Fig. 87c [57a]; note the many unknown compounds (cf. also reference [468]).

Tancredi and Curtius [422] investigated the ether-soluble DNP-amino acids from the urine of healthy and sick children. Several chromatograms with different loads were obtained for each urine specimen. The results are shown in Table 52.

Usually, a volume corresponding to $10^{-4} \times 24$ hr. quantity or its multiple is applied from the ether-soluble fraction. Fig. 88 shows the

Table 51. Detectability[1] of the ether-soluble dinitrophenyl compounds in urine (see Fig. 86) [309]. $+$ = detectable; $-$ = not detectable; v = masked.

DNP	μl Urine				DNP	μl Urine			
	40	80	160	320		40	80	160	320
Alanine	+	+	+	v	X_1	−	−	−	+
β-alanine	−	+	+	+	X_2	−	−	−	+
α-aminoadipic acid	−	−	+	+	X_3	−	−	−	+
α-aminobutyric acid	−	−	+	+	X_4	−	−	−	+
β-aminoisobutyric acid	+	+	+	v	X_5	−	−	−	+
Asparagine	+	v	v	v	X_6	−	−	−	+
Aspartic acid	−	−	−	+	X_7	−	−	−	+
Cystine	+	v	v	v	X_8	−	−	+	+
Glutamine	+	v	v	v	X_9	−	−	−	+
Glutamic acid	+	+	+	+	X_{10}	−	−	−	+
Glycine	+	+	+	+	X_{11}	−	+	+	+
Histidine	−	−	+	v	X_{12}	−	+	+	+
Isoleucine	+	+	v	v	X_{13}	−	−	−	+
Leucine	+	+	v	v	X_{14}	+	+	+	+
Lysine + Tyrosine[2]	+	+	+	+	X_{15}	+	+	+	+
Methionine sulfone	−	−	+	+	X_{16}	−	−	−	+
Ornithine	−	+	+	+	X_{17}	−	−	−	+
Phenylalanine	+	+	+	+	X_{18}	+	+	+	v
Proline	−	−	−	v[2]	X_{19}	+	+	+	v
Serine	+	+	+	+	Number of				
Threonine	+	+	+	+	detectable				
Tryptophan	−	−	−	v[2]	DNP-				
Valine	−	+	+	+	compounds	19	22	26	30
A_1[3]	+	+	+	+					
A_2[3]	−	+	+	+					

[1] 24-hr. urine; detection in UV-light.
[2] Detected by rechromatography (see text).
[3] Artifact.

ether-soluble DNP-amino acids from the urine of a 58 year old female with an unknown metabolic disorder; it can be recognized that cystine + asparagine (unseparated), glutamine, serine, glycine, and β-aminoisobutyric acid appear in unusually large amounts compared to the other amino acids [306]. Inborn errors of protein metabolism often can be distinguished from other metabolic disorders by stress tests. Thin-layer chromatography offers a rapid orientation concerning the variations in amino acid excretion in such cases. Figs. 88b and c show the variation of amino acid excretion after a lysine stress (Colombo and Pataki [70]).

a

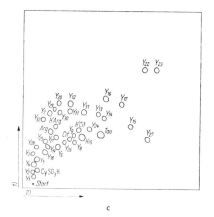

c

a) Water-soluble DNP-amino acids from 50 μl morning urine (see text) according to Fig. 87b "UV-photography" [306], cf. p. 14.

b) Spot pattern of the water-soluble DNP-amino acids on Silica Gel G layers; first dimension: V = pyridine, VI = n-butanol at room temperature with 25% saturated ammonia. 36 = CySO₃H, 34 = Arg, 35 = Cit, 37 = Tau, 15 = Di-His. From Brenner et al. [434].

c) Water-soluble DNP-compounds from urine on Silica Gel HF₂₅₄ layers (30 g silica gel + 65-70 ml water for preparation of the layer). 1 = phenol/water/17% ammonia (80 : 20 : 2); 2 = 2-chloroethane/toluene/pyridine/25% ammonia (50 : 35 : 15 : 10). Intermediate drying: overnight at room temperature and 10 min. at 60°C H'Cit = homocitrulline, H'Arg = homoarginine. From Bürgi et al. [57a].

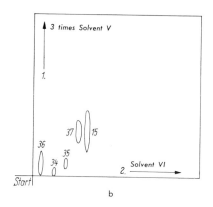

b

Figs. 87a, b, c. Thin-layer chromatographic detection of the water-soluble DNP-amino acids.

It is remarkable that not only the lysine but among others also the α-aminoadipic acid concentration is elevated. Further information on stress tests can be found in Bürgi et al. [57a].

Goedde et al. [53a] investigated the quantitative determination of β-aminoisobutyric acid in urine. Chromatography was carried out on Silica Gel G layers with n-heptane/pyridine/glacial acetic acid (50 : 50 : 0.1, developed three times) or with pyridine/chloroform/n-heptane/glacial acetic acid (50 : 25 : 25 : 1) and after drying (3 hrs. in vacuum), with chloroform/glacial acetic acid (100 : 1); iterative; two or three developments, see p. 16) in the same direction. A β-AiB-solution is applied as a control. The layer is scraped off and β-AiB is

Table 52. Ether-soluble dinitrophenylamino acids from children's urine [422]. The data refer to applied quantities corresponding to μg total nitrogen.

Amino acid	A[1]			B[2]				C[3]			D[4]	
	15	37.5	75	5	15	37.5	75	15	37.5	75	15	75
Glycine	+	+	+	+[5]	+[5]	+[5]	+[5]	−	+	+	+	+
Glutamine	+	+	+	+[6]	+	+	+	−	−	−	+	+
Asparagine	+	+	+	−	+	+	+	−	+	+	+	+
Alanine	−	+	+	−	−	+	+	−	+	+	+	+
Histidine[7]	−	+	+	−	+	+	+	−	−	−	−	−
Tyrosine/lysine	−	−	+[6]	−	−	+	+	−	−	−	+[5]	+[5]
Serine	−	−	−	−	+	+	+	−	−	−	−	−
Cystine	−	−	−	−	+	+	+	−	−	+	−	+
Tryptophan	−	−	−	−	−	+	+	−	−	−	−	+
α-aminobutyric acid	−	−	−	−	−	−	+	−	−	−	−	−
Proline	−	−	−	−	−	−	+	−	−	−	+	+
Glutamic acid	−	−	−	−	−	−	+	−	−	−	−	−
Threonine	−	−	−	−	−	+	+	−	−	+	−	+
Cysteine	−	−	−	−	−	−	−	−	−	+	−	−
Valine	−	−	−	−	−	−	−	−	−	−	+	+
Leucin/Isoleucine	−	−	−	−	−	−	−	−	−	−	+	+
X[8]	−	−	−	−	−	+	+	−	−	−	−	−

[1] Normal girl, 3 years old.
[2] Hyperglycineme, 5-year-old boy.
[3] Hypoaminoaciduria in a 16-month-old child with anorexia.
[4] Renal insufficiency, 7-year-old girl.
[5] Large quantity.
[6] Trace.
[7] DNP-histidine was determined only in the ether-soluble fraction.
[8] Unknown spot between start and $(CyS)_2$ (compare Fig. 86).

determined at 366 mμ after elution (6 ml 10% acetic acid; shake for 1 hr.). See [53a] and literature listed in this reference concerning the importance of β-AiB-excretion.

We detected 21 amino acids in 60 μl serum (Pataki and Keller [311]). The results are shown in Table 50 (chromatography of the ether- and water-soluble DNP-amino acids according to above data). Our findings are in good agreement with literature; in the case of cysteic acid and methioninesulfone *in vitro oxidation products* are probably formed. In contrast to the literature data [339], we were unable to detect histidine; this result is in agreement with the thin-layer chromatographic findings of Opienska-Blauth (see Fig. 81a).

Human ejaculate is known to contain several proteolytic enzymes (Mann [240]). Krampitz and Doepfmer [210] were able to detect a considerable leucine aminopeptidase activity; recently we [202] found a high "oxytocinase" activity. By thin-layer chromatography, it was possible to detect 24 amino acids in sperm fluid, which in our opinion give a picture of the proteolytic activity (Table 50) [196]. The separation of ether-soluble DNP-amino acids was carried out according to the method

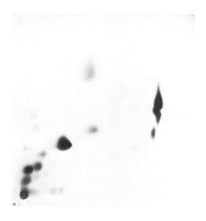
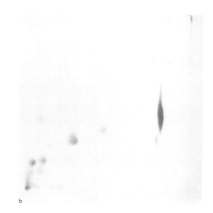

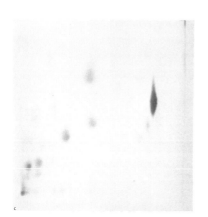

Fig. 88a. Ether-soluble DNP-amino acids from the urine of a 58-year old patient with unknown metabolic disorder (compare Fig. 85a); note that cystine + asparagine (unseparated), glutamine, serine, glycine and β-aminoisobutyric acid appear in unusually large amounts [306].

Fig. 88b and c. Ether-soluble DNP-amino acids from the urine of a child with normal metabolism before (b) and after (c) lysine stress (compare Fig. 85a).

Note that not only the lysine but also the α-aminoadipic acid concentration increases under a lysine stress [70].

on p. 137 (Fig. 89a); a good separation is possible with a two-dimensional chromatogram [196]. Development of the water-soluble DNP-derivatives was carried out in one dimension on Silica Gel G layers with n-butanol saturated with 25% ammonia at room temperature (Fig. 89b) [196].

2. Other Applications

Drawert et al. [100] isolated C^{14}-amino acids as well as C^{14}-alcohols from fermentation lots of grape must containing wine yeast and C^{14}-glutamic acid. The method of Brenner et al. (p. 137, method A) served for the chromatography of the ether-soluble DNP-derivatives. The separation can be seen in Fig. 90. The counts corrected for the background are also shown (for the radioactivity determination, see p. 35).

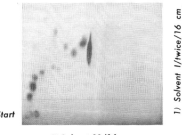

Fig. 89a. Ether-soluble DNP-amino acids from human ejaculate (compare Fig. 74a); load corresponding to 50 μl sperm; in 100 μl sperm, 4 additional spots (see Table 50) can be seen—photography in normal light (p. 142). From Keller and Pataki [196].

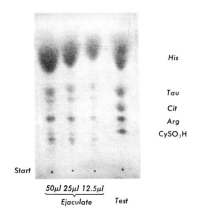

Fig. 89b. Water-soluble DNP-amino acids from human ejaculate; daylight photography (p. 142). Solvent: n-butanol at room temperature saturated with 25% ammonia. From Keller and Pataki [196].

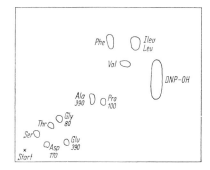

Fig. 90. Ether-soluble DNP-amino acids from fermentation charges of grape must with wine yeast and C^{14}-glutamic acid; the data refer to counts corrected for the background. Experimental conditions according to Fig. 73a. From Drawert et al. [100].

Palmork [292] converted the amino acids present in seawater with DNFB, isolated the ether- and butanol-soluble fractions, and used circular chromatography for separation (p. 15, Silica Gel G; benzyl alcohol/glacial acetic acid/chloroform 30 : 3 : 70). The ether phase gave 12 zones, of which 4 could be identified (DNP-derivatives of glycine, threonine, valine, and phenylalanine); 3 zones could be recognized in the butanol phase.

References

[1] Abderhalden, E., and W. Stix: Z. Physiol. Chem. **129**, 143 (1923).

[2] Akabori, S., T. Ikenaka, Y. Okada and K. Kohno: Proc. Imp. Acad. Japan **29**, 509 (1953); cited acc. to [6].

[3] Akabori, S., K. Ohno and K. Narita: Bull. Chem. Soc. Japan **25**, 214 (1952).

[4] Antener, I.: Exper. Med. Surg. **12**, 57 (1954).

[5] Arx, E. von, and R. Neher: J. Chromatog. **12**, 329 (1963).

[6] Bailey, J. L.: Techniques in Protein Chemistry, Elsevier, Amsterdam (1962).

[7] Bancher, E., H. Scherz and K. Kaindl: Mikrochim. Acta 652 (1964).

[8] Bancher, E., H. Scherz and V. Prey: Mikrochim. Acta 712 (1963).

[9] Barbier, M., H. Jäger, H. Tobias and E. Wyss: Helv. Chim. Acta **42**, 2440 (1959).

[10] Baron, D. N., and J. Economidis: J. Clin. Pathol. **16**, 484 (1963).

[11] Barrolier, J.: Naturwissenschaften **48**, 404 (1961).

[12] Barrolier, J.: Naturwissenschaften **48**, 554 (1961).

[13] Barrolier, J. (Schering AG, Berlin): Personal communication (1964).

[14] Barrolier, J., and J. Heilmann: Z. Physiol. Chem. **304**, 21 (1956).

[15] Barrolier, J., J. Heilmann and E. Watzke: Z. Physiol. Chem. **309**, 219 (1957).

[16] Barry, J. M.: Biochem. J. **63**, 669 (1956).

[17] Baudet, P., Cl. Otten and E. Cherbuliez: Helv. Chim. Acta **47**, 2430 (1964).

[18] Baxter, C. F., and I. Senonen: Anal. Biochem. **7**, 55 (1964).

[19] Beale, D., and J. K. Whitehead: Tritium in the Physical and Biological Sciences [Vienna] **1**, 179 (1962).

[20]   Behrens, M., and W. Neu: Biochim. Biophys. Acta **78**, 540 (1963).

[21]   Behrens, U., M. Ringpfeil and G. Striegler: Monatsber. Deut. Akad. Wiss. Berlin **3**, 635 (1961).

[21a] Bekeresky, J.: Anal. Chem. **35**, 261 (1963).

[22]   Bennet, R. D., and E. Heftmann: J. Chromatog. **12**, 245 (1963).

[23]   Berger, H.: Schweiz. Med. Wochschr. **83**, 33, 761 (1953); **86**, 711, 729 (1956).

[24]   Berger, H.: Aminoaciduria and Hyperaminoaciduria, Karger-Verlag, Basel (1959).

[25]   Berger, H., I. Antener, T. Brechbühler and G. Stalder: Ann. Paediat. (Basel) **202**, 465 (1964).

[26]   Berger, J. A., G. Meynel, J. Petit and P. Blanquet: Bull. Soc. Chim. France 2662 (1963).

[27]   Bergmann, E. D., and M. Bentov: J. Org. Chem. **26**, 1480 (1961).

[28]   Bernath, G., and K. Koczka: Acta. Chim. Acad. Sci. Hung. **31**, 443 (1962).

[29]   Bhandari, P. R., B. Lerch and G. Wohlleben: Pharm. Ztg. Ver. Apotheker-Ztg. **107**, 1618 (1962).

[30]   Bigwood, E. J., R. Crocart, E. Schramm, P. Soupart and H. Vis: Advan. Clin. Chem. **2**, 201 (1959).

[31]   Biserte, G., J. W. Hollemann, J. Hollemann-Dehove and P. Sautiere: Chromatog. Rev. **2**, 59 (1960).

[32]   Biserte, G., M. Paysant, F. Ferraz and M. Halpern: Arquiv. Port. Bioquim. **3**, 360 (1953); see Chem. Abstr. **56**, 1703c (1962).

[33]   Biserte, G., T. Plaquet-Schonaert, P. Boulanger and P. Paysant: J. Chromatog. **3**, 25 (1960).

[34]   Boissonas, R. A., and S. Lo Bianco: Experientia **8**, 425 (1952).

[35]   Boissonas, R. A., St. Guttmann, R. L. Huguenin, P. A. Jaquenaud and Ed. Sandrin: Helv. Chim. Acta **46**, 2347 (1963).

[36]   Bolliger, H. R. (Hoffmann La Roche AG, Basel): Personal communication (1964).

[37]   Bonker, G. J., and B. L. Tonge: J. Chromatog. **12**, 52 (1963).

[38]   Borecky, J., J. Gasparic and M. Vecera: Collection Czech. Chem. Commun. **24**, 1822 (1959).

[39]   Brandner, G., and A. J. Virtanen: Acta Chem. Scand. **17**, 2563 (1963).

[40]   Brenner, M., and W. Hofer: Helv. Chim. Acta **44**, 1794 (1961).

[41]   Brenner, M., H. R. Müller and E. Lichtenberg: Helv. Chim. Acta **35**, 217 (1952).

[42]   Brenner, M., and A. Niederwieser: Experientia **16**, 378 (1960).

[43]   Brenner, M., and A. Niederwieser: Experientia **17**, 237 (1961).

[44]   Brenner, M., A. Niederwieser and G. Pataki: Experientia **17**, 145 (1961).

[45]   Brenner, M., A. Niederwieser and G. Pataki: in A. T. James and L. J. Morris, New Biochemical Separations, Van Nostrand Co., London (1964).

[46]   Brenner, M., A. Niederwieser and G. Pataki: in E. Stahl, Thin-layer Chromatography, Academic Press, New York, 1965.

[47]   Brenner, M., A. Niederwieser, G. Pataki and A. R. Fahmy: Experientia **18**, 101 (1962).

[48]   Brenner, M., A. Niederwieser, G. Pataki and R. Weber: in E. Stahl, Thin-Layer Chromatography, Academic Press, New York, 1965.

[49]   Brenner, M., and G. Pataki: Helv. Chim. Acta **44**, 1420 (1961).

[50]   Brenner, M., G. Pataki and A. Niederwieser: in G. B. Marini-Bettolo, Thin Layer Chromatography, Elsevier, Amsterdam (1964) p. 116.

[51]   Brenner, M. The Stationary Phase in Paper- and Thin-Layer Chromatography, Elsevier Amsterdam (1965), pp. 322 ff.

[52]   Brieskorn, C. H., and J. Glasz: Naturwissenschaften **51**, 216 (1964).

[53]   Brodasky, T. F.: Anal. Chem. **36**, 996 (1964).

[53a] Brunschede, H., R. Hofbauer and H. W. Goedde: Klin. Wochschr. **43**, 93 (1965); Z. Anal. Chem. **212**, 191 (1965).

[54]   Bujard, El. and J. Mauron, J. Chromatog. **21**, 19 (1966).

[55]   Bujard, El. (Afico AG, La-Tour-de-Peilz, Schweiz): Personal communication (1964).

[56]   Bungenberg de Jong, H. G., and J. Th. Hoogeveen: Proc. Acad. Sci. Amsterdam **63B**, 228, 243, 383 (1960).

[57]   Bungenberg de Jong, H. G., and J. Th. Hoogeveen: Proc. Acad. Sci. Amsterdam **64B**, 1, 18, 167, 183 (1961).

[57a] Bürgi, W., J. P. Colombo and R. Richterich: Klin. Wochschr., **43**, 1202 (1965).

[58]   Bush, I. E.: The Chromatography of Steroids, Pergamon Press, London (1961).

[59]   Caldin, D. J. Mc.: Chem. Rev. **60**, 39 (1960).

[60]   Carisano, A.: J. Chromatog. **13**, 83 (1964).

[61]   Cerny, V., J. Joska and L. Labler: Collection Czech. Chem. Commun. **26**, 1658 (1961).

[62]   Checchi, I., de Richardi: Bull. Soc. Ital. Farm. Ospital **7**, 165 (1961).

[63]   Cherbuliez, E., Br. Baehler, J. Marszalek, A. R. Sussmann and J. Rabinowitz: Helv. Chim. Acta **46**, 2446 (1963).

[64] Cherbuliez, E., Br. Baehler and J. Rabinowitz: Helv. Chim. Acta **43**, 1871 (1960).

[65] Cherbuliez, E., Br. Baehler and J. Rabinowitz: Helv. Chim. Acta **47**, 1350 (1964).

[66] Cherbuliez, E., J. Marszalek and J. Rabinowitz: Helv. Chim. Acta **46**, 1445 (1963).
      43, 87 (1960).

[67] Cherbuliez, E., A. R. Sussmann and J. Rabinowitz: Pharm. Acta Helv. **36**, 131 (1961).

[68] Chinball, A. C., and M. W. Rees: The Chemical Structure of Proteins, Little Brown & Co., Boston (1952) p. 70.

[69] Codern, L., J. Gadea, E. Gal and M. Montagut: Afinidad [Barcelona] **20**, 159 (1963).

[70] Colombo, J. P., and G. Pataki: Unpublished.

[71] Connors, W. M., and W. K. Boak: J. Chromatog. **16**, 243 (1964).

[72] Consden, R., A. H. Gordon and A. J. P. Martin: Biochem. J. **40**, 33 (1946).

[73] Consden, R., A. H. Gordon and A. J. P. Martin: Biochem. J. **41**, 590 (1947).

[74] Crawhall, J. C., E. Saunders and C. J. Thompson: Federation European Biochem. Societies, First Meeting, London, 23-25 March 1964, Abstr. p. 91.

[75] Crove, M. O. L.: Anal. Chem. **13**, 845 (1941).

[76] Cseh, G. (Institut für Organische Chemie der Universität Basel): Personal communication (1964).

[77] Curtius, H. Ch.: Dissertation University of Basel (1961).

[78] Dahn, H., and H. Fuchs: Helv. Chim. Acta **45**, 261 (1962).

[79] Dauvillier, P.: J. Chromatog. **11**, 405 (1963).

[80] David, I. B., T. C. French and J. M. Buchanan: J. Biol. Chem. **238**, 2178 (1963).

[81] David, I. B., T. C. French and J. M. Buchanan (Depart. of Chemistry, Massachusetts Institute of Technology): Personal communication (1964).

[82] Davidek, J., and J. Blattna: J. Chromatog. **7**, 204 (1962).

[83] Davidek, J., and E. Davidkova: Pharmazie **16**, 352 (1961).

[84] Davies, B. H.: J. Chromatog. **10**, 518 (1963).

[85] Decker, P.: Naturwissenschaften **44**, 305 (1957).

[86] Decker, P.: Naturwissenschaften **45**, 464 (1958).

[87] Decker, P., and K. Kikodse: Z. Anal. Chem. **165**, 337 (1959).

[88] Decker, P., W. Riffart and G. Oberneder: Naturwissenschaften **38**, 288 (1951).

[89] Demole, E.: J. Chromatog. **1**, 30 (1958).

[90]   Dent, C. E.: Biochem. J. **43**, 169 (1948).
[91]   Dent, C. E.: Brit. Med. Bull. **10**, 247 (1954).
[92]   Determann, H.: Gel Chromatography, Springer, Berlin-New York, 1967.
[93]   Determann, H., K. Bohnhard, R. Köhler and Th. Wieland: Helv. Chim. Acta **46**, 2498 (1963).
[94]   Dhont, J. H., and C. de Roy: Analyst **86**, 74 (1961).
[95]   Dhont, J. H., and C. de Roy: Analyst **86**, 527 (1961).
[96]   Dimillier, I., and R. G. Trout: J. Thoracic Cardiovascular Surgery, St. Louis, **48**, 822 (1964).
[97]   Dittmann, J.: Z. Klin. Chem. **1**, 190 (1963).
[98]   Dittmann, J.: Z. Klin. Chem. **3**, 59 (1965).
[99]   Djerassi, C., K. Undheim, R. C. Sheppard, G. W. Terry and B. Sjöberg: Acta Chem. Scand. **15**, 903 (1961).
[100] Drawert, F., O. Bachmann and K. H. Reuther: J. Chromatog. **9**, 376 (1962).
[101] Duncan, G. R.: J. Chromatog. **8**, 37 (1962).
[102] Edman, P.: Acta Chem. Scand. **4**, 277, 283 (1950).
[103] Edman, P.: Acta Chem. Scand. **10**, 761 (1956).
[104] Edman, P.: Ann. N.Y. Acad. Sci. **85**, 602 (1960).
[105] Edwards, C. H., E. L. Gadsden, L. B. Carter and G. Edwards: J. Chromatog. **2**, 188 (1959).
[106] Ehrhardt, E., and F. Cramer: J. Chromatog. **7**, 405 (1962).
[107] Eriksson, S., and J. Sjöquist: Biochem. Biophys. Acta **45**, 290 (1960).
[108] Euler, H. von, H. Hasselquist and I. Limnell: Arkiv Kemi **21**, 259 (1963).
[109] Fahmy, A. R., A. Niederwieser, G. Pataki and M. Brenner: Helv. Chim. Acta **44**, 2022 (1961).
[110] Fasella, P., C. Turano, A. Giartosio: in G. B. Marini-Bettolo, Thin Layer Chromatography, Elsevier, Amsterdam (1964) p. 205.
[111] Faucenott, L., and M. Waldesbühl: Pharm. Acta Helv. **38**, 423 (1963).
[112] Feltkamp, H.: Deut. Apotheker-Ztg. **102**, 1269 (1962).
[113] Filippovich, Yu. B.: Zh. Analit. Khim. **15**, 374 (1960); see Chem. Abstr. **55**, 9531h (1961).
[114] Fittkau, S., H. Hansen, I. Marquardt, H. Diessner and U. Kettmann: Z. Physiol. Chem. **338**, 180 (1964).
[115] Flodin, P.: Dissertation Universität Uppsala (1962) available from A.B. Pharmacia, Uppsala, Sweden.
[116] Floss, H. G., and D. Gröger: Z. Naturforsch. **18B**, 519 (1963).
[117] Floyd, N. F., M. S. Cammaroti and T. F. Lavine: Arch. Biochem. Biophys. **102**, 343 (1963).

[118] Fraenkel-Conrat, H., and J. I. Harris: J. Am. Chem. Soc. **76,** 6058 (1954).

[119] Fraenkel-Conrat, H., J. I. Harris and A. L. Levy: Methods Biochem. Anal. **2,** 393 (1955).

[120] Franc, J.: Collection Czech. Chem. Comm. **24,** 2102 (1959).

[121] Franc, J., and J. Jokl: Collection Czech. Chem. Comm. **21,** 1161 (1956).

[122] Franc, J., and J. Jokl: J. Chromatog. **2,** 423 (1959).

[123] Franc, J., and J. Latinak: Chem. Listy **49,** 317 (1955).

[124] Franc, J., and J. Latinak: Chem Listy **49,** 325 (1955).

[125] Franc, J., and J. Latinak: Chem. Listy **49,** 328 (1955).

[126] Fray, G., and J. Frey: Bull. Soc. Chim. Biol. **45,** 1201 (1963).

[127] Frodyma, M., R. W. Frei and D. J. Williams: J. Chromatog. **13,** 61 (1964).

[128] Frodyma, M., and R. W. Frei: J. Chromatog. **15,** 501 (1964).

[128a] Frodyma, M., and R. W. Frei: J. Chromatog. **17,** 131 (1965).

[129] Galanos, D. S., and V. M. Kapoulas: J. Chromatog. **13,** 128 (1964).

[130] Gamp, A., P. Studer, H. Linde and Kuno Meyer: Experientia **18,** 292 (1962).

[131] Gasior, E., M. Pietrusiewicz, H. Kowalska and J. Opienska-Blauth: Acta Biochim. Polon. **5,** 333 (1958).

[132] Gasparic, J., and M. Vecera: Collection Czech. Chem. Comm. **22,** 1426 (1957).

[133] Geiss, F., and H. Schlitt: Naturwissenschaften **50,** 350 (1963).

[134] Geiss, F., H. Schlitt, F. J. Ritter and M. Weimar: J. Chromatog. **12,** 469 (1963).

[135] Gelotte, B.: in A. T. James and L. J. Morris, New Biochemical Separations, Van Nostrand Co., London (1964) p. 94.

[136] Gerlach, H., J. A. Owtschinnikow and V. Prelog: Helv. Chim. Acta **47,** 2294 (1964).

[137] Giddings, J. C., G. H. Steward and A. L. Ruoff: J. Chromatog. **3,** 239 (1960).

[138] Glasz, J.: Dissertation Universität Würzburg (1964).

[139] Gorchein, A.: Biochim. Biophys. Acta **84,** 356 (1964).

[140] Gray, W. R., and B. S. Hartley: Biochem. J. **89,** 59P (1963).

[141] Grebenovsky, E.: Z. Anal. Chem. **185,** 290 (1962).

[142] Green, J., S. Marcinkiewicz: J. Chromatog. **10,** 35 (1963).

[143] Green, J., S. Marcinkiewicz and D. McHale: J. Chromatog. **10,** 158 (1963).

[144] Gros, C., M. Private de Garilhe, A. Costopanagiotis and R. Schwyzer: Helv. Chim. Acta **44,** 2042 (1961).

[145] Gut, V., and J. Rudinger: Collection Czech. Chem. Comm. **28,** 2593 (1963).

[146] Guttmann, St., and R. A. Boissonas: Helv. Chim. Acta **46**, 1626 (1963).

[147] Guttmann, St., J. Pless and R. A. Boissonas: Helv. Chim. Acta **45**, 170 (1962).

[148] Hagermann, D. D., and J. M. Spencer: Steroids **4**, 547 (1964).

[149] Halmekoski, J., and H. Hannikainen: Suomen Kemistilehti **36B**, 24 (1963).

[150] Halpaap, H.: Chem.-Ing.-Tech. **35**, 488 (1963).

[151] Hannig, K.: Unpublished (communication of Desaga GmbH, Heidelberg).

[152] Hara, S. H., Tanaka, and M. Takeuchi: Chem. Pharm. Bull. (Tokyo) **12**, 626 (1964).

[153] Harris, C. K., E. Tigane and G. S. Hanes: Can. J. Biochem. Physiol. **39**, 439 (1961).

[154] Harris, J. I., and P. Roos: Biochem. J. **71**, 434 (1959).

[155] Hartley, B. S., and V. Massey: Biochm. Biophys. Acta **21**, 58 (1956).

[156] Havinga, E., C. Schattenkerk, G. Heymens-Visser, and K. E. T. Kerling: Rec. Trav. Chim. **83**, 672 (1964).

[157] Heesing, A.: Chem. Ber. **96**, 2170 (1963).

[158] Hesse, G., and M. Alexander: Immediate Separation and Chromatography (1961), cited according to Wohlleben [445].

[158a] Heyns K., and W. Walter: Naturwissenschaften **39**, 507 (1952).

[159] Hill, R. L., and W. R. Schmidt: J. Biol. Chem. **237**, 389 (1962).

[160] Hirs, C. H. W.: J. Biol. Chem. **219**, 611 (1956).

[161] Hofmann, A. F.: Biochem. Biophys. Acta **60**, 458 (1962).

[162] Hollingsworth, D. R., M. Dillard, and P. K. Bondy: J. Lab. Clin. Med. **62**, 346 (1963).

[163] Honegger, C. G.: Helv. Chim. Acta **44**, 173 (1961).

[164] Honegger, C. G.: Helv. Chim. Acta **45**, 1409 (1962).

[165] Honegger, C. G.: Helv. Chim. Acta **46**, 1730 (1963).

[166] Honegger, C. G.: Helv. Chim. Acta **46**, 1772 (1963).

[167] Honegger, C. G.: Helv. Chim. Acta **47**, 2384 (1964).

[168] Honegger, C. G.: The Stationary Phase in Chromatography, Elsevier Amsterdam (1965) p. 318.

[169] Hörhammer, L., H. Wagner, and G. Bittner: Der Deutsche Apotheker **14**, 148 (1962).

[170] Hörhammer, L., H. Wagner, and F. Kilger: Der Deutsche Apotheker **15**, 164 (1963).

[171] Hromatka, O., and W. A. Aue: Monatsh. Chem. **95**, 503 (1962).

[172] Hromatka, O. (Institut für Organische Chemie der TH Wien): Personal communication (1963).

[173] Huber, J., W. Schaknies and J. Rückbeil: Pharmazie **18**, 37 (1963).

[174] Huguenin, R. L., and R. A. Boissonas: Helv. Chim. Acta **44**, 213 (1961).

[175] Ilse, D., and P. Edman: Australian J. Chem. **16**, 411 (1963).

[176] Ingram, V. M.: J. Chem. Soc. 193 (1953).

[177] Ingram, V. M.: Biochim. Biophys. Acta **28**, 539 (1958).

[178] Ishii, S. J., and B. Witkop: J. Am. Chem. Soc. **85**, 1832 (1963).

[179] Ishii, S. J., and B. Witkop: J. Am. Chem. Soc. **86**, 1848 (1964).

[180] Ismailov, N. A., and M. S. Shraiber: Farmatsija [Sofia] 1 (1938).

[181] Jaquenod, P.A., and R. A. Boissonas: Helv. Chim. Acta **45**, 1462 (1962).

[182] Jatzkewitz, H., and E. Mehl: Z. Physiol. Chem. **32**, 251 (1960).

[183] Jänchen, D.: J. Chromatog. **14**, 261 (1964).

[184] Jeanes, A., C. S. Wise and R. J. Dimler: Anal. Chem. **23**, 415 (1951).

[185] Johansson, B. G., and L. Rymo: Acta Chem. Scand. **16**, 2067 (1962).

[186] Jost, K., J. Rudinger, and F. Sorm: Collection Czech. Chem. Comm. **26**, 2496 (1961).

[187] Jutisz, M., and P. de la Llosa: Bull. Soc. Chim. France, 2913 (1963).

[188] Kaplan, J. M., and Schneider, F. L.: Microchem. J. **6**, 557 (1962).

[189] Kappeler, H., and R. Schwyzer: Helv. Chim. Acta **44**, 1136 (1961).

[190] Karpitschka, N.: Mikrochim. Acta 157 (1963).

[191] Katz, A. M., W. J. Dreyer and C. B. Anfinsen: J. Biol. Chem. **234**, 2897 (1957).

[192] Kaufmann, H. P., H. Wessels and C. Bondopadhyaya: Fette, Seifen, Anstrichmittel **65**, 543 (1963).

[193] Keil, B.: Collection Czech. Chem. Comm. **23**, 740 (1958).

[194] Keil, B., V. Tomasek and J. Sedlachova: Chem. Listy **46**, 457 (1952).

[195] Kelemen, J., and G. Pataki: Z. Anal. Chem. **195**, 81 (1963).

[196] Keller, M., and G. Pataki: Helv. Chim. Acta **46**, 1687 (1963).

[197] Keller-Schierlein, W., and A. Deer: Helv. Chim. Acta **46**, 1907 (1963).

[198] Keulemans, A. J. M.: Gas Chromatography, Reinhold Publ. Co., New York (1957).

[199] Kirchner, J. G., J. M. Miller and G. J. Keller: Anal. Chem. **23**, 420 (1953).

[200] Klaus, R.: J. Chromatog. **16**, 311 (1964).

[201] Klee, W., and M. Brenner: Helv. Chim. Acta **44**, 2151 (1964).

[202] Klimek, R., M. Keller and G. Pataki: Unpublished.

[203] Kobor, J., G. Bernath, and K. Koczka: Acta Physica Chimica [Szegedin, Hungary] 7, 121 (1961).

[204] Kobor, J., G. Bernath, and K. Koczka: Pedagogiai Föiskola Evkönyve [Yearbook of the College of Higher Education, Szegedin, Hungary], 167 (1961).

[205] Kofranyi, E.: Z. Physiol. Chem. 299, 129 (1955).

[206] Konigsberg, W., and R. J. Hill: J. Biol. Chem. 237, 2547 (1962).

[207] Korff, R. W. von, B. Pollara, R. Coyne, J. Runquist, and R. Kapoor: Biochim. Biophys. Acta 74, 698 (1963).

[208] Korzun, B. P., L. Dorfmann, and S. M. Brody: Anal. Chem. 35, 950 (1963).

[209] Kowkabany, G. M., and H. G. Cassidy: Anal. Chem. 22, 817 (1950).

[210] Krampitz, G., and R. Doepfmer: Klin. Wochschr. 39, 1300 (1961).

[211] Kratzl, K., and G. Puschmann: Holzforschung 14, 1 (1960).

[212] Kraut, H., and H. Zimmermann-Telschow: Nutr. Dieta 4, 22 (1962).

[213] Kündig, W., and N. Neukom: Helv. Chim. Acta 46, 1423 (1963).

[214] Lagoni, H., and A. Wortmann: Milchwissenschaft 10, 360 (1955).

[215] Lagoni, H., and A. Wortmann: Milchwissenschaft 11, 206 (1956).

[216] Larsson, H.: J. Chromatog. 11, 331 (1963).

[217] Laver, W. G.: Biochem. Biophys. Acta 53, 469 (1961).

[218] Lawrence, L., and W. J. Moore: J. Am. Chem. Soc. 73, 3973 (1951).

[219] Lederer, M.: Proceedings Second. Internat. Congr. Surface Activity, Butterworths, London (1957) p. 506.

[220] Lees, T. M., and P. J. De Muria: J. Chromatog. 8, 108 (1962).

[221] Leibnitz, E., U. Behrens, and M. Ringpfeil: Wasserwirtsch. Wassertech. 6, 299 (1956); 7, 3 (1957).

[222] Lenk, H. P.: Z. Anal. Chem. 184, 107 (1961).

[223] Levy, A. L., and D. Chung: J. Am. Chem. Soc. 77, 2899 (1955).

[224] Levy, A. L., and D. Chung: Biochim. Biophys. Acta 17, 454 (1955).

[225] Levy, A. L., and C. H. Li: J. Biol. Chem. 213, 487 (1955).

[226] Lisboa, B. P.: J. Chromatog. 13, 391 (1964).

[227] Lisboa, B. P., and E. Diczfalusy: Acta Endocrinol. 40, 60 (1962).

[228] Llosa, P. de la, C. Tertrin, and M. Jutisz: Biochim. Biophys. Acta **93**, 40 (1964).
[229] Llosa, P. de la, C. Tertrin, and M. Jutisz: Experientia **20**, 204 (1964).
[230] Llosa, P. de la, C. Tertrin, and M. Jutisz: J. Chromatog. **14**, 136 (1964).
[231] Lockhart, I. M., and E. P. Abraham: Biochem. J. **58**, 633 (1954).
[232] Lucas, F. J., T. B. Shaw, and S. G. Smith: Anal. Biochem. **6**, 335 (1963).
[233] Lüdy-Tenger, F.: Pharm. Acta Helv. **37**, 770 (1962).
[234] Macek, K.: Chem. Listy **48**, 1181 (1954).
[235] Macek, K.: Experientia **17**, 162 (1961).
[236] Macek, K.: Magy. Kem. Lapja [Budapest] 297 (1962).
[237] Macek, K., and Z. Vejdelek: Nature **176**, 1173 (1955).
[238] Mangold, H. K.: J. Am. Oil. Chemists Soc. **38**, 708 (1961).
[239] Mangold, H. K.: In E. Stahl, Thin-layer Chromatography, Academic Press, New York, 1965, p. 58.
[240] Mann, T.: The Biochemistry of Semen, Methuen & Co., London (1954).
[241] Marcinkiewicz, S., and J. Green: J. Chromatog. **10**, 184 (1963).
[242] Marcinkiewicz, S., J. Green, and D. McHall: J. Chromatog. **10**, 42 (1963).
[243] Marcucci, F., and E. Mussini: J. Chromatog. **11**, 270 (1963).
[244] Marquart, P., H. Schmidt, and M. Späth: Arzneimittel-Forsch. **13**, 1100 (1963).
[245] Martin, A. J. P.: Biochem. Soc. Symp. (Cambridge, England) **3**, 4 (1950).
[246] Martin, E. C.: J. Chem. Soc. 3935 (1961).
[247] Massaglia, A., and U. Rosa: J. Chromatog. **14**, 516 (1964).
[248] Matthias, W.: Naturwissenschaften **41**, 17 (1954).
[249] Mazur, R. H., B. W. Ellis, and P. S. Cammarata: J. Biol. Chem. **237**, 1619 (1962).
[250] Meinhard, I. E., and N. F. Hall: Anal. Chem. **21**, 158 (1949).
[251] Michl, H., and H. Bachmeyer: Monatsh. Chem. **95**, 480 (1964).
[252] Miller, J. M., and J. G. Kirchner: Anal. Chem. **26**, 2002 (1954).
[253] Milne, M. D.: Brit. Med. J. 327 (1964).
[254] Moffat, E. D., and R. L. Lyttle: Anal. Chem. **31**, 926 (1959).
[255] Moghissi, A.: Anal. Chim. Acta **30**, 91 (1964).
[256] Monder, C.: Biochem. J. **90**, 522 (1964).
[257] Montant, C., and I. M. Tourze-Poulet: Bull. Soc. Chim. Biol. **42**, 161 (1960).
[258] Moore, S.: J. Biol. Chem. **238**, 235 (1963).
[259] Morris, C. J. O. R.: J. Chromatog. **16**, 167 (1964).

[260] Moses, V.: J. Chromatog. **9,** 241 (1962).
[261] Mottier, M.: Mitt. Lebensmittelunters. Hyg. [Bern] **49,** 454 (1958).
[262] Mottier, M., and M. Potterat: Anal. Chim. Acta **13,** 46 (1955).
[263] Müller, R. H., and D. L. Clegg: Anal. Chem. **23,** 269 (1951).
[264] Munnier, R.: J. Chromatog. **1,** 524 (1958).
[265] Mussini, E., and F. Marcucci: J. Chromatog. **17,** 576 (1965).
[266] Mussini, E., and F. Marcucci (Istituto di Richerche Farmacologiche "Mario Negri," Milan): Personal communication (1964).
[267] Mutschler, E., and H. Rochelmeyer: Arch. Pharm. **292,** 449 (1959).
[268] Müting, D.: Naturwissenschaften **39,** 303 (1952).
[269] Müting, D., and E. Kaiser: Z. Physiol. Chem. **332,** 276 (1963).
[270] Myhill, D., and D. S. Jackson: Anal. Biochem. **6,** 193 (1963).
[271] Neu, W.: Diplomarbeit, Universität Giessen (1963).
[272] Niederwieser, A.: Dissertation, Universität Basel (1962).
[273] Niederwieser, A., and M. Brenner: Experientia **21,** 50, 105 (1965).
[274] Niederwieser, A., and G. Pataki: Chimia [Aarau] **14,** 378 (1960).
[275] Niedrich, H.: Chem. Ber. **97,** 2527 (1964).
[276] Niemann, A.: Naturwissenschaften **47,** 514 (1960).
[277] Nischwitz, E.: Z. Anal. Chem. **193,** 190 (1963).
[278] Nomoto, M., Y. Narahashi, and H. Murakami: J. Biochem. [Tokyo] **48,** 906 (1960).
[279] Nürnberg, E.: Arch. Pharm. **292,** 610 (1959).
[280] Nybom, N.: J. Chromatog. **14,** 120 (1964).
[281] Nybom, N.: Physiologia Plantarum [Kobenhavn] **17,** 434 (1964).
[282] Oepen, H., and I. Oepen: Klin. Wochschr. **41,** 921 (1963).
[283] Offer, G. W.: Biochem. Biophys. Acta **90,** 193 (1964).
[284] Opienska-Blauth, J.: Clin. Chim. Acta **4,** 841 (1959).
[284] Opienska-Blauth: Personal communication (1963); cited acc. to Pataki [305].
[286] Opienska-Blauth, J., M. Charezinski, and H. Brebec: Anal. Biochem. **6,** 69 (1963).
[287] Opienska-Blauth, M. Charezinski, M. Sanecka, and H. Bruszkiewicz: J. Chromatog. **7,** 321 (1962).
[288] Opienska-Blauth, J., and A. Gebala: Biochemical Clinics [New York] 205 (1963).
[289] Opienska-Blauth, J., H. Kraczkowski, and H. Bruszkiewicz: In G-B. Marini-Bettolo, Thin Layer Chromatography, Elsevier, Amsterdam (1964).

[290] Oswald, N., and H. Flück: Pharm. Acta Helv. **39**, 293 (1964).
[291] Palmer, K. H.: Can. Pharm. J., Sci. Sect. **96**, 250 (1963).
[292] Palmork, K. H.: Acta Chem. Scand. **17**, 1456 (1963).
[293] Pataki, G.: Chimia [Aarau] **18**, 23 (1964).
[294] Pataki, G.: Chimia [Aarau] **18**, 24 (1964).
[295] Pataki, G.: Dissertation, Universität Basel (1962).
[296] Pataki, G.: Ergebnisse. Labor. Med. **2**, 163 (1965).
[297] Pataki, G.: Helv. Chim. Acta **47**, 784 (1964).
[298] Pataki, G.: Helv. Chim. Acta **47**, 1763 (1964).
[299] Pataki, G.: J. Chromatog. **12**, 541 (1963).
[300] Pataki, G.: J. Chromatog. **16**, 541 (1964).
[300a] Pataki, G., and A. Kunz: Unpublished.
[301] Pataki, G.: J. Chromatog. **16**, 553 (1964).
[302] Pataki, G.: J. Chromatog. **17**, 580 (1965).
[303] Pataki, G.: J. Chromatog. **17**, 327 (1965).
[304] Pataki, G.: Schweiz. Med. Wochschr. **94**, 1789 (1964).
[305] Pataki, G.: Z. Klin. Chem. **2**, 129 (1964).
[306] Pataki, G.: Unpublished.
[307] Pataki, G., and J. Kelemen: J. Chromatog. **11**, 50 (1963).
[308] Pataki, G., and M. Keller: Helv. Chim. Acta **46**, 1054 (1963).
[309] Pataki, G., and M. Keller: Helv. Chim. Acta **47**, 787 (1964).
[310] Pataki, G., and M. Keller: Klin. Wochschr. **43**, 227 (1965).
[311] Pataki, G., and M. Keller: Z. Klin. Chem. **1**, 157 (1963).
[312] Pastuska, G., and H. J. Petrowitz: Chemiker Ztg. **86**, 311 (1962).
[313] Pastuska, G., and H. Trinks: Chemiker Ztg. **85**, 535 (1961).
[314] Pastuska, G., and H. Trinks: Chemiker Ztg. **86**, 135 (1962).
[315] Patterson, S. J., and R. L. Clements: Analyst **89**, 328 (1964).
[316] Paulson, J. C., F. E. Deatherage, and E. F. Almy: J. Am. Chem. Soc. **75**, 2039 (1953).
[317] Pawlowa, L. B.: Vopr. Vitaminol. Altaisk. Gos. Med. Inst. 77 (1959), see Chem. Abstr. **55**, 22464h (1961).
[318] Peraino, C., and A. E. Harper: Anal. Chem. **33**, 1863 (1961).
[319] Peter, H., M. Brugger, J. Schreiber, and A. Eschenmoser: Helv. Chim. Acta **46**, 577 (1963).
[320] Petschik, H., and E. Steger: J. Chromatog. **9**, 307 (1962).
[321] Pikkarainen, J., and E. Kulonen: Ann. Med. Exp. Biol. Fenniae **37**, 382 (1959).
[322] Plattner, Pl. A., K. Vogler, R. O. Studer, P. Quitt, and W. Keller-Schierlein: Helv. Chim. Acta **46**, 927 (1963).
[323] Poethke, W., and W. Kinze: Pharm. Zentralhalle **101**, 685 (1962).

[324] Pöhm, M.: Naturwissenschaften **48,** 551 (1961).

[325] Pollara, B., and R. von Korff: Biochim. Biophys. Acta **39,** 364 (1960).

[326] Pravda, Z., K. Poduska, and K. Blaha: Collection Czech. Chem. Comm. **29,** 2626 (1964).

[327] Purdy, S. J., and E. V. Truter: Analyst **87,** 802 (1962).

[328] Purdy, S. J., and E. V. Truter: Chem. Ind. 506 (1962).

[329] Purdy, S. J., and E. V. Truter: Lab. Pract. **13,** 500 (1964).

[330] Rachinskii, V. V.: In I. M. Hais and K. Macek, Some General Problems of Paper Chromatography, Academy of Sci. [Prague] (1962).

[331] Ramachandran, L. K., A. Epp, and G. McConnell: Anal. Chem. **27,** 1734 (1955).

[332] Rao, K. R., and H. R. Sober: J. Am. Chem. Soc. **76,** 1328 (1954).

[333] Rastekiene, L., and T. Prauskiene: Lietuvos T.S.R. Mosklu Akad. Darbai Ser. B. 5 (1963); see Chem. Abstr. **60,** 6217h (1964).

[334] Ratney, R. S.: J. Chromatog. **11,** 111 (1963).

[335] Ratney, R. S. (Hood College, Frederick, Md.): Personal communication (1964).

[336] Reichl, E. R.: Mikrochim. Acta 955 (1956).

[337] Reitsema, R. H.: Anal. Chem. **26,** 960 (1954).

[338] Relvas, M. A.: Clinica Chim. Acta **8,** 12 (1963).

[339] Relvas, M. A., and F. S. Ferraz: Clinica Chim. Acta **8,** 533 (1963).

[340] Riniker, B.: Cited acc. to E. Stahl, Thin-Layer Chromatography, Springer Verlag, Berlin (1962), p. 424.

[341] Riniker, B., and R. Schwyzer: Helv. Chim. Acta **47,** 2357 (1964).

[342] Riniker, B., and R. Schwyzer: Helv. Chim. Acta **47,** 2375 (1964).

[343] Ritschard, W. J.: J. Chromatog. **16,** 327 (1964).

[344] Rittel, W.: Helv. Chim. Acta **45,** 2465 (1962).

[345] Ritter, F. J., and G. M. Meyer: Nature **193,** 941 (1962).

[346] Rokkones, T.: Scand. J. Clin. Lab. Invest. **16,** 149 (1964).

[347] Rokkones, T. (Central Laboratory, Ulleval Hospital, Oslo): Personal communication (1965).

[348] Ronkainnen, P.: J. Chromatog. **11,** 228 (1963).

[349] Rudinger, J., H. Farkasova: Collection Czech. Chem. Comm. **28,** 2941 (1963).

[350] Ruoff, A. L., and J. C. Giddings: J. Chromatog. **3,** 438 (1960).

[351] Russel, D. W.: Biochem. J. **87,** 1 (1963).

[352] Russel, D. W.: Biochem. J. **83**, 8P (1962).

[353] Russel, D. W.: J. Chem. Soc. 894 (1963).

[354] Rybicka, S. M.: Chem. Ind. 308 (1962).

[355] Sanger, F.: Biochem. J. **39**, 507 (1945).

[356] Sanger, F.: J. Polymer. Sci. **49**, 3 (1961).

[357] Sanger, F., and E. O. P. Thompson: Biochim. Biophys. Acta **71**, 468 (1963).

[358] Sanger, F., and E. O. P. Thompson: Biochem. J. **53**, 353 (1953).

[359] Sarges, R., and B. Witkop: J. Am. Chem. Soc. **86**, 1862 (1964).

[360] Satake, K., T. Okuyama, M. Ohashi and T. Shinoda: J. Biochem. [Tokyo] **47**, 654 (1960).

[361] Schauer, H. K., and R. Bulirsch: Z. Naturforsch. **13B**, 327 (1958).

[362] Schellenberg, P.: Angew. Chem. **74**, 118 (1962).

[363] Schildknecht, H., and O. Volkert: Naturwissenschaften **50**, 442 (1963).

[364] Schilling, E. D., P. J. Burchill and R. A. Clayton: Anal. Biochem. **5**, 1 (1963).

[365] Schlicher, H.: Z. Analyt. Chem. **199**, 335 (1963).

[366] Schneider, G., and C. Schneider: Z. Physiol. Chem. **332**, 316 (1963).

[367] Schreier, K.: Congenital Metabolic Anomalies, Thieme Verlag, Stuttgart (1963).

[368] Schreier, K.: Z. Anal. Chem. **201**, 103 (1964).

[369] Schroeder, W. A., J. R. Shelton, J. Balog-Shelton and J. Cormick: Biochemistry **2**, 992 (1963).

[370] Schröder, E.: Liebig's Ann. **673**, 186 (1964).

[371] Schröder, E.: Liebig's Ann. **673**, 220 (1964).

[372] Schröder, E. (Schering AG, Berlin): Personal communication (1964).

[373] Schröder, E., and H. Gibian: Ann. Chem. **673**, 176 (1964).

[374] Schulze, P. E., and M. Wenzel: Angew. Chem. **74**, 777 (1962).

[375] Schwyzer, R., A. Costopanagiotis and P. Sieber: Helv. Chim. Acta **46**, 870 (1963).

[376] Schwyzer, R., and H. Dietrich: Helv. Chim. Acta **44**, 2003 (1961).

[377] Schwyzer, R., B. Iseli, H. Kappeler, B. Riniker, W. Rittel and H. Zuber: Helv. Chim. Acta **46**, 1975 (1963).

[378] Schwyzer, R., and H. Kappeler: Helv. Chim. Acta **46**, 1550 (1963).

[379] Schwyzer, R., B. Riniker and H. Kappeler: Helv. Chim. Acta **46**, 1541 (1963).

[380] Schwyzer, R., W. Rittel and A. Costopanagiotis: Helv. Chim. Acta **45**, 2473 (1962).

[381] Schwyzer, R., and P. Sieber: Nature **199**, 172 (1963).

[382] Seher, A.: Fette, Seifen, Anstrichmittel **61**, 345 (1959).

[383] Seiler, N., G. Werner and H. Wiechmann: Naturwissenschaften **50**, 643 (1963).

[384] Seiler, N., and H. Wiechmann: Experientia **20**, 559 (1964).

[385] Seiler, N., and H. Wiechmann: Z. Physiol. Chem. **337**, 229 (1964).

[386] Shasha, B., and R. L. Whistler: J. Chromatog. **14**, 532 (1964).

[387] Shellard, E. J.: Lab. Pract. **13**, 290 (1964).

[388] Sheppard, H., and W. H. Tsien: Anal. Chem. **35**, 1992 (1963).

[389] Siepmann, E., and H. Zahn: Biochim. Biophys. Acta **82**, 412 (1964).

[390] Sjöholm, I.: Acta Chem. Scand. **18**, 889 (1964).

[391] Sjöquist, J.: Arkiv Kemi **11**, 129 (1957).

[392] Sjöquist, J.: Arkiv Kemi **14**, 291 (1959).

[393] Sjöquist, J.: Biochim. Biophys. Acta **41**, 20 (1960).

[394] Sjöquist, J., B. Blombäck and P. Wallen: Arkiv Kemi **16**, 425 (1961).

[395] Skarzynski, B., and M. Sarnecka-Keller: Advan. Clin. Chem. **5**, 107 (1962).

[396] Smith, I.: Chromatographic and Electrophoretic Techniques, Interscience Publ. Inc., New York (1962).

[397] Smyth, D. G., and D. F. Elliott: Analyst **89**, 81 (1964).

[398] Smyth, D. G., W. H. Stein and S. Moore: J. Biol. Chem. **238**, 227 (1963).

[399] Snyder, F., and N. Stephens: Anal. Biochem. **4**, 128 (1962).

[400] Squibb, R. L.: Nature **198**, 317 (1963).

[401] Squibb, R. L.: Nature **199**, 1216 (1963).

[402] Stahl, E.: Arch. Pharm. **292**, 411 (1959).

[403] Stahl, E.: Arch. Pharm. **293**, 531 (1960).

[404] Stahl, E.: Chem.-Ingr.-Tech. **36**, 941 (1964).

[405] Stahl, E.: Chemiker Ztg. **82**, 323 (1958).

[406] Stahl, E.: Lab. Pract. **13**, 496 (1964).

[407] Stahl, E.: Parfüm. and Kosmetik **39**, 564 (1958).

[408] Stahl, E.: Pharm. Rundschau 1 (1959).

[409] Stahl, E.: Pharmazie **11**, 633 (1956).

[410] Stahl, E.: Thin-layer Chromatography, Academic Press, New York (1965).

[411] Stahl, E., and U. Kaltenbach: J. Chromatog. **5**, 531 (1961).

[412] Stahl, E., and J. Pfeifle: Z. Anal. Chem. **200**, 377 (1964).

[413]   Stegemann, H., and B. Lerch (Institut für Biochemie der Biologischen Bundesanstalt, Hann. Münden, Germany): Personal communication (1965).

[413a] Stegemann, H., R. Hillebrecht and W. Rien: Z. Physiol. Chem. 340, 11 (1965).

[414]   Stegemann, H., and B. Lerch: Anal. Biochem. 9, 417 (1964).

[415]   Stein, W. H., and S. Moore: J. Biol. Chem. 190, 103 (1951).

[416]   Steven, F. S.: Anal. Biochem. 4, 316 (1962).

[417]   Steven, F. S.: J. Chromatog. 8, 417 (1962).

[418]   Steward, J. M., and D. W. Wooley: Biochemistry 3, 700 (1964).

[419]   Studer, R. O.: Helv. Chim. Acta 46, 421 (1963).

[420]   Studer, R. O., W. Lergier and K. Vogler: Helv. Chim. Acta 46, 612 (1963).

[421]   Studer, R. O., K. Vogler and W. Lergier: Helv. Chim. Acta 44, 131 (1961).

[422]   Tancredi, F., and H. C. Curtius (Universitäts-Kinderklinik, Zurich): Personal communication (1964).

[423]   Thoma, J. A.: Anal. Chem. 35, 214 (1963).

[424]   Tonge, B. L.: Nature 195, 491 (1962).

[425]   Tower, D. B., E. L. Peters and J. R. Wherett: J. Biol. Chem. 237, 1861 (1962).

[426]   Truter, E. V.: J. Chromatog. 14, 57 (1964).

[427]   Turba, F., and G. Grundlach: Biochem. Z. 326, 322 (1955).

[428]   Türler, M., and O. Högl: Mitt. Lebensmittelunters. Hyg. [Bern] 52, 123 (1961).

[429]   Underwood, C. E., and F. E. Deatherage: Science 115, 95 (1952).

[430]   Vogler, K., R. O. Studer, P. Lanz, W. Lergier, E. Böhni and B. Fust: Helv. Chim. Acta 46, 2823 (1963).

[431]   Vogler, K., R. O. Studer and W. Lergier: Helv. Chim. Acta 44, 1495 (1961).

[432]   Vogler, K., R. O. Studer, W. Lergier and P. Lanz: Helv. Chim. Acta 43, 1751 (1960).

[433]   Wallenfels, K., and A. Arens: Biochem. Z. 332, 217 (1960).

[434]   Walz, D., A. R. Fahmy, G. Pataki, A. Niederwieser and M. Brenner: Experientia 19, 213 (1963).

[435]   Wasicky, R.: Naturwissenschaften 50, 569 (1963).

[436]   Weitzel, G., S. Hörnle and F. Schneider: Ann. Chem. 677, 190 (1964).

[436a] Wenzel, M.: Naturwissenschaften 52, 129 (1965).

[437]   Wenzel, M., and P. E. Schulze: Z. Analyt. Chem. 201, 349 (1964).

[438] Wieland, Th., and H. Bende: Chem. Ber. **98**, 504 (1965).

[439] Wieland, Th., and U. Gebert: Anal. Biochem. **6**, 201 (1963).

[440] Wieland, Th., and D. Georgopoulos: Biochem. Z. **340**, 476 (1964).

[441] Wieland, Th., D. Georgopoulos and H. Kampe: Biochem. Z. **340**, 483 (1964).

[442] Wieland, Th., G. Lüben and H. Determann: Experientia **18**, 430 (1962).

[443] Williams, T. J.: Introduction to Chromatography, Blackie & Sons, Glasgow (1947).

[444] Witkop, B. (National Institutes of Health, Bethesda, Maryland): Personal communication (1964).

[445] Wohlleben, G.: G. I. T. Fachzeitschrift für Laboratorium **7**, 1, 43, 221 (1963).

[446] Wohlleben, G. (Woelm AG, Eschwege, Germany): Personal communication (1964).

[447] Wollenweber, P.: J. Chromatog. **9**, 369 (1962).

[448] Wollish, E. G., M. Schmall and M. Hawrylyshyn: Anal. Chem. **33**, 1138 (1961).

[449] Wood, S. E., and H. H. Strain: Anal. Chem. **26**, 260 (1954).

[450] Wünsch, E., H. G. Heidrich and W. Grassmann: Chem. Ber. **97**, 1818 (1964).

[451] Zacharius, R. M., and E. A. Tailley: Anal. Chem. **34**, 1551 (1962).

[452] Zacharius, R. M., and E. A. Tailley: J. Chromatog. **7**, 51 (1962).

[453] Zahn, H., and H. Pfannmüller: Biochem. Z. **330**, 97 (1958).

[454] Zahn, H., and E. Siepmann: Biochem. Z. **335**, 303 (1961).

[455] Zbiral, E.: Monatsh. Chem. **94**, 639 (1963).

[456] Zöllner, N., G. Wolfram and G. Arnim: Klin. Wochschr. **40**, 273 (1962).

[457] Zuber, H.: Chimia [Aarau] **14**, 405 (1960).

ADDENDUM

[458] Ambert, J., C. Pechery and C. Carpentier, Ann. Biol. Clin. (Paris) **24**, 17 (1966).

[459] Bielesky, R. L., and N. A. Turner: Anal. Biochem. **17**, 278 (1966).

[460] Cole, M., J. C. Fletcher and A. Robson, J.: Chromatog. **20**, 616 (1965).

[461] Deyl, Z., and J. Rosmus, J.: Chromatog. **20**, 514 (1965).

[462] Figge, K.: Clin. Chim. Acta **12**, ˜5 (1965).

[463] Frei, R. W., I. T. Fukui, V. T. Lieu and M. M. Frodyma: Chimia [Aarau] **20**, 23 (1966).

[464] Glaesmer, R., K. Ruckpaul and W. Jung: Z. Med. Labortech. **6**, 175 (1965).

[465] Habermann, E.: Naunyn-Schmiedebergs Arch. Exp. Pathol. Pharmakol. **253**, 474 (1966).

[466] Heatchcot, J. G., and K. Jones: J. Chromatog. **24**, 106 (1966).

[467] Holbrook, J. J., G. Pfleiderer, J. Schnetger and S. Diemair: Biochem. Z. **344**, 1 (1966).

[468] Keller, M., and G. Pataki: Klin. Wochenschr. **44**, 99 (1966).

[469] Morse, D., and B. L. Horecker: Anal. Biochem. **14**, 429 (1966).

[470] Mucha, S., and H. Ochi: Bunseki Kagaku **14**, 728 (1965); thru Chem. Abstr. **63**, 15525E (1965).

[471] Munier, R. L., and G. Sarrazin: Bull. Soc. Chim. France 2959 (1965).

[472] Munier, R. L., and G. Sarrazin: Bull. Soc. Chim. France, 1490 and 2427 (1965); 1363, 1365 and 1367 (1966).

[473] Munier, R. L., and G. Sarrazin: J. Chromatog. **22**, 347 (1966).

[474] Niederwieser, A.: J. Chromatog., in press.

[475] Niederwieser, A.: personal communication (1966).

[476] Niederwieser, A., and C. G. Honegger: in C. Giddings and R. A. Keller (editors), Advances in Chromatography, Vol. 2.

[477] Pataki, G., and A. Niederwieser: J. Chromatog., **29**, 133 (1967).

[478] Pataki, G., and Ed. Strasky: Chimia [Aarau] **20**, 361 (1966).

[479] Rosetti, V.: Biochem. Appl. [Parma] **11**, 225 (1964).

[480] Seiler, N., and H. Wiechmann: Z. Analyt. Chemie. **220**, 109 (1966).

[481] Tinelli, R.: Bull. Soc. Chim. Biol. **47**, 182 (1966).

[482] Troughton, W. D., R. C. Brown and N. A. Turner: Am. J. Clin. Pathol. **46**, 139 (1966).

[483] Turner, N. A., and R. J. Redgwell: J. Chromatog. **21**, 129 (1966).

[484] Wang, K. T., J. M. K. Huang and I. S. Y. Wang: J. Chromatog. **22**, 362 (1966).

[485] Wang, K. T., J. M. K. Huang: J. Chromatog. **24**, 460 (1966).

[486] Habermann, E.: unpublished.

[487] Pataki, G.: Chromatogr. Reviews, **9**, 23 (1967).

[488] Woods, K. R., and K. T. Wang: Biochim. Biophys. Acta **133**, 369 (1967).

[489] Edman, P., and G. Begg: Europ. J. Biochem. **1**, 80 (1967).

Bibliography of
New Applications

(A compilation of the most important papers)
Numbers refer to following literature citations

CROSS INDEX

Free Amino Acids—490, 491, 493, 494, 528, 534, 537, 545, 546, 554, 555, 562, 564, 567, 575, 583, 584, 592, 596, 613, 614, 643, 648, 655, 669, 678, 679, 680, 689, 693, 701, 708, 710, 716, 732, 741, 751, 759, 764, 767, 769, 778, 779, 783, 785, 795, 801, 809, 811, 812, 813, 814, 815, 816, 843, 849, 855, 859, 879, 880, 886, 909, 917, 926, 933, 934, 937, 949, 971, 972.

Iodo Amino Acids—558, 559, 597, 601, 615, 649, 652, 758, 765, 770, 797, 805, 821, 840, 856, 857, 870, 878, 892, 953, 975, 976, 977, 978.

Peptides and Intermediates of Peptide Synthesis—509, 510, 511, 527, 531, 538, 548, 557, 563, 565, 566, 567, 572, 574, 590, 600, 603, 607, 608, 609, 610, 616, 629, 630, 632, 635, 641, 644, 668, 670, 671, 672, 673, 683, 691, 697, 698, 702, 703, 704, 705, 725, 752, 777, 787, 827, 828, 834, 867, 874, 875, 876, 889, 894, 896, 899, 916, 928, 929, 930, 931, 932, 940, 941, 942, 948, 964.

DNP-Amino Acids—512, 540, 560, 570, 574, 621, 623, 653, 657, 674, 690, 733, 808, 809, 810, 811, 812, 813, 840, 846, 901, 910, 918, 944, 945, 946.

DNAP-Amino Acids—580, 838.

DNS-Amino Acids—501, 631, 654, 699, 725, 762, 780, 788, 808, 810, 811, 813, 837, 866, 867, 877, 899, 966.

PTH-Amino Acids—590, 640, 685, 694, 748, 782, 788, 807, 808, 810, 811, 813, 915, 947, 959.

Combination with Spectroscopy and Fluorometry—512, 613, 808, 810, 811, 813, 972.

Nitropyridyl-, Dinitropyridyl- and Nitropyrimidyl-Amino Acids —517, 518, 553.

Fingerprint Technique—548, 567, 668, 725, 787, 867.

Benzothiadiazine Derivatives of Amino Acids—551, 552.

"Pipsyl"-Amino Acids—561.

Amino Acid-Copper-Complexes—559.

Trinitrobenzenesul-phonyl-Amino Acids—794.

Dinitrosulphophenyl-Amino Acids—829.

Nitropyrididylium-N-oxide Derivatives of Amino Acids—884.

Methylthiohydantoin Derivatives of Amino Acids—890, 891.

Miscellaneous Applications (Including Sequence Analysis)— 514, 522, 523, 530, 540, 569, 594, 595, 606, 694, 737, 811.

REFERENCES

490—Affonso, A.: "Chromatography of amino acids on plaster of Paris," J. Chromatog. **22**, 452 (1966).

491—Affonso, A.: "Electrophoresis on plaster of Paris," J. Chromatog. **31**, 646 (1967).

492—Aloof-Hirsch, S., A. de Vries and A. Berger: "The direct lytic factor of cobra venom: purification and chemical characterization," Biochim. Biophys. Acta **154**, 53 (1968).

493—Ambert, J. P., C. Pechery, A. Lemonnier and L. Hartmann: "Estimation of urinary amino acids II. In cirrhotic patients before and after portocaval anastanosis," Ann. Biol. Clin. [Paris] **24**, 41 (1966).

494—Ambert, J. P., C. Pechery, P. Moukhtar, E. Housset and L. Hartmann: "Estimation of urinary amino acids III. In the normal dog and in the dog after termino-lateral or latero-lateral portocaval anastanosis," Ann. Biol. Clin. [Paris] **24,** 51 (1966).

495—Anastasi, A., V. Erspamer and R. Endean: "Isolation and structure of caerulein, an active decapeptide from the skin of hyla caerulea," Experientia **23,** 699 (1967).

496—Andersen, S. O.: "Covalent cross-links in a structural protein: resilin," Acta Physiol. Scand. **66,** Suppl. **263** (1966).

497—Anderson, J. C., M. A. Barton, P. M. Hardy, G. W. Kenner, J. Preston and R. C. Sheppard: "Peptides XXIV," J. Chem. Soc. **1967,** 108.

498—Antonovics, I., and G. T. Young: "Amino acids and peptides XXV. The mechanism of the base-catalyzed racemization of the p-nitrophenyl esters of the acyl peptides," J. Chem. Soc. **1967,** 595.

499—Arakawa, K., M. Nakatani and M. Nakamura: "Purification of human angiotensin," Nature **214,** 278 (1967).

500—Argoudelis, A. D., R. R. Herr, D. J. Mason, T. R. Pyke and J. F. Zieserl: "New amino acids from streptomyces," Biochemistry **6,** 165 (1967).

501—Arnott, M. S., and D. N. Ward: "Separation of dansyl amino acids in a single analysis," Anal. Biochem. **21,** 50 (1967).

502—Aurich, A., H. P. Kleber and W. D. Schöpp: "An inducible carnitine dehydrogenase from Pseudomonas aeruginosa," Biochim. Biophys. Acta **139,** 505 (1967).

503—Bachmayer, H., and H. Michl: "Hämolytisch wirksame Stoffe in Molchgiften," Monatsh. Chem. **96,** 1166 (1965).

504—Bajusz, S.: "Contribution to the discussion on the protecting groups of arginine," Acta Chim. Acad. Sci. Hung. **44,** 31 (1965).

505—Bajusz, S., and T. Lazar: "Synthesis of peptides related to the C-terminal 25-39 sequences of corticotropins," Acta Chim. Acad. Sci. Hung. **48,** 111 (1966).

506—Bajusz, S., K. Medzihradszky, Z. Paulay and Z. Lang: "Totalsynthese des menschlichen Corticotropins," Acta Chim. Acad. Sci. Hung. **52,** 335 (1967).

507—Balabanova-Radonova, E., I. Kluh, J. Vancek and F. Sorm: "On proteins XCVII," Collection Czech. Chem. Comm. **30,** 2241 (1965).

508—Barrett, G. C.: "Cleavage of N-thiobenzoyl-dipeptides with TFA: the basis of a new stepwise degradation of polypeptides," Collection Czech. Chem. Comm. **1967,** 487.

509—Barth, A.: "Ueber den Nachweis des Aminolyseverlaufes von Cbo-Aminosäure-4-(phenylazo)-phenylestern mit Hilfe der DC," J. Prakt. Chem. **27,** 181 (1965).

510—Barth, A., and P. Schwenk: "Ueber Aminosäure-phenylazo-phenyl-Derivate V," J. Prakt. Chem. (4) 32, 130 (1966).

511—Barth, A., and R. Wierick: "Spektralphotometrische Untersuchungen der Aminolyse von N-geschützten Aminosäure-4-(phenylazo)-phenylestern," J. Prakt. Chem. 33, 61 (1966).

512—Baudet, P., C. Otten and K. Eder: "Nouveaux aspects de micro-méthodes pour la détermination des structures on chimie organique," Arch. des Sci. [Genève] 18, 287 (1965).

513—Bayer, S. H., P. Hathaway, F. Pascasio, J. Bordley and Ch. Orton: "Differences in the amino acid sequences of tryptic peptides from three sheep hemoglobin beta-chains," J. Biol. Chem. 242, 2211 (1965).

514—Beacham, J., P. H. Bentley, R. A. Gregory, G. W. Kenner, J. K. McLeod and R. C. Sheppard: "Synthesis of human gastrin I," Nature 209, 585 (1966).

515—Belenkij, B. G., E. S. Gankina, S. A. Pryanishnikova and D. P. Erastou: Mol. Biol. (2) 1, 184 (1967).

516—Belenkij, B. G., E. S. Gankina and V. V. Nesterov: Dokl. Akad. Nauk S.S.S.R. 172, 91 (1967).

517—Bello, C. di, and A. Signor: "TLC of dinitropyridyl- and nitro-pyrimidyl-amino acids," J. Chromatog. 17, 506 (1965).

518—Bello, C. di, and A. Signor: "Cromatografia su strato sottile di dinitropiridil- e nitropiridil-ammino-acidi," Chim. Ind. Ital. 47, 92 (1965).

519—Benöhr, H. C., F. Frisch: "Carboxylesterase aus Rinderlebermikrosomen I," Z. Physiol. Chem. 348, 1102 (1967).

520—Benson, A. M., H. F. Mower and K. T. Yasunobu: "The amino acid sequence of clostridium butyricum ferredoxin," Arch. Biochem. Biophysics 121, 563 (1967).

521—Bentley, P. H., G. W. Kenner and R. C. Sheppard: "Structures of human gastrin I and II," Nature 209, 583 (1966).

522—Bentley, P. H., and J. S. Morley: "Polypeptides II: A 'hydrazino-peptide' analogue of norophthalmic acid amide," J. Chem. Soc. 1966, 60.

523—Berg, T. L., L. O. Froholm and S. G. Laland: "The biosynthesis of gramicidin S in a cellfree system," Biochem. J. 96, 43 (1965).

524—Bernhammer, E., and K. Krisch: "Zur Hydrolyse von Aminosäure-Arylamiden durch mikrosomale Schweineleberesterase und Serum," Z. Klin. Chem. 4, 49 (1966).

525—Bhagavan, N. V., P. M. Rao, L. W. Pollard, R. K. Rao, T. Winnick and J. B. Hall: "The biosynthesis of gramicidin S. A rest study," Biochemistry 5, 3844 (1966).

526—Bishop, D. G., L. Rutberg and B. Samuelsson: "The chemical composition of the cytoplasmic membrane of bacillus subtilis," Europ. J. Biochem. **2**, 448 (1967).

527—Blaha, K., and J. Rudinger: "Stereoisomeric tri- and hexapeptides containing glycine, phenylalanine and leucine," Collection Czech. Chem. Comm. **32**, 2365 (1967).

528—Blanc, P., P. Bertrand, G. de Sagni-Sannes and R. Lescure: "CCM de quelques acides amines," Chim. Anal. [Paris] **47**, 283 (1965).

529—Bleich, R.: "Analyse eines Hydrolysats von Plodia- und Ephestiapinnfäden durch DC," Naturwissenschaften **54**, 170 (1967).

530—Blombäck, B., M. Blombäck, P. Edman and B. Hessel: "Human fibrinopeptides, isolation and characterization," Biochim. Biophys. Acta **115**, 371 (1966).

531—Bloom, S. M., S. K. Dasgupta, R. P. Patel and E. R. Blout: "The synthesis of Gly-L-Pro-Gly and Gly-L-Pro-Ala oligopeptides and sequential polypeptides," J. Am. Chem. Soc. **88**, 2035 (1966).

532—Bobst, A., and M. Viscontini: "Hydroxylation nonenzymatique de la phénylalanine en tyrosine à l'aide de ptèrines tetrahydrogénées," Helv. Chim. Acta **49**, 884 (1966).

533—Bøhmer, T., and J. Bremer: "Propionylcarnitine in animal tissue," Biochim. Biophys. Acta **152**, 440 (1968).

534—Bondivenne, R., and N. Busch: "Séparation et dosage d'acides aminés en chromatographie sur couches mines," J. Chromatog. **29**, 349 (1967).

535—Bornstein, P.: "Comparative sequence studies of rat skin and tendon collagen. I. Evidence for incomplete hydroxylation of individual prolyl residues in the normal proteins," Biochemistry **6**, 3082 (1967).

536—Bownds, D.: "Site of attachment of retinal in rhodopsin," Nature **216**, 1178 (1967).

537—Breinlich, J.: "Beitrag zur Analytik von Aminosäurenmischungen zur parenteralen Eiweissernährung," Pharm. Ztg. (50) **111**, 1866 (1966).

538—Brenner, M., H. C. Curtius and M. Kny: "Aminoacyleinlagerung 5. Mitt," Helv. Chim. Acta **49**, 250 (1966).

539—Bretzel, G.: "Ueber Thynnin: Fraktionierung und Isolierung einer einheitlichen Komponente," Z. Physiol. Chem. **348**, 419 (1967).

540—Brias, E., J. M. Ghuysen and P. Dezelee: "The cell wall peptidoglycan of bacillus megaterium K. M. I. Studies on the stereochemistry of a,a'-diaminopimelic acid," Biochemistry **6**, 2598 (1967).

541—Bricteux-Gregoire, A., R. Schyns and M. Florkin: "Structure des peptides libérés au cours de l'activation du trypsinogène de mouton," Biochim. Biophys. Acta. **127**, 227 (1966).

542—Broen, E. G., and A. V. Silver: "The natural occurrence of a uracil-5-peptide and its metabolic relationship to GMP-5′," Biochim. Biophys. Acta 119, 1 (1966).

543—Brown, J. L., S. Koorajian, J. Katze and I. Zabin: "Beta-Galactosidase: amino- and carboxy-terminal studies," J. Biol. Chem. 241, 2826 (1966).

544—Bruns, F. H., H. Reinauer and W. Stork: "Analytische und biologische Studien über den Gehalt an N-Acetyl-L-aspartat im Gehirn," Z. Physiol. Chem. 348, 512 (1967).

545—Bujard, E.: "Séparation bidimensionelle des acides aminés basiques, neutres et acides par chromatographie en couche mince sur cellulose," Chrom. Symp. III Bruxelles, Presses Académiques Européennes, 1964, p. 145.

546—Bujard, E.: "Séparation bidimensionelle des acides aminés basiques, neutres et acides par ccm sur cellulose," J. Pharm. Belg. 20, 413 (1965).

547—Bumans, R., A. Veiss and M. Olte: "Crystalline aluminium oxide as adsorbent in TLC," Latvijas PSR Zinatnu Akad. Vestis. Kim. Ser. 1965, 35 (1965).

548—Burns, D. J., and N. A. Turner: "Peptide mapping on cellulose thin layers," J. Chromatog. 30, 469 (1967).

549—Burns, V. W., and D. L. Wong: "Chromatographic separation on silica gel glass fiber paper of compounds from yeast labeled by aspartate-^{14}C," J. Chromatog. 26, 542 (1967).

550—Camejo, G.: "Structural studies of rat plasma lipoproteins," Biochemistry 6, 3228 (1967).

551—Cameroni, R., M. T. Bernabei, M. Facchini and V. Ferioli: "Recerche nel campo della 1,2,4-benzotiadiazina XXXVIII. Reazione fra 3-(2-cloroetil)-7-nitro-1,2,4-benzotiadiazin-1,1-diossido e dipeptide," Farmaco [Pavia] Ed. Sci. 22, 37 (1967).

552—Cameroni, R., M. T. Bernabei, V. Ferioli and A. Albasini: "Su alcune proprietà dei benzotiadiazin-derivati di aminoacidi," Gazz. Chim. Ital. 95, 786 (1965).

553—Celon, E., L. Biondi and E. Bordignon: "Detection and quantitative micro-determination of nitro-2-pyridyl amino acids by TLC on precoated sheets," J. Chromatog. 35, 47 (1968).

554—Chan, Y. K., and J. P. Riley: "Determination of amino acids in sea water," Deep-Sea Res. 13, 1115 (1966).

555—Chiari, D., M. Röhr and G. Widtmann: "Die Anwendung von Zellulosepulvern zur dc-Trennung von Aminosäuren," Mikrochim. Acta (4) 1965, 669.

556—Chillemi, F.: "Sintesi di peptide-idrazidi contenenti un residuo dell'acido beta-t-butilaspartico," Gazz. Chim. Ital. 96, 359 (1966).

557—Chimiak, A., and J. Rudinger: "Amino acids and peptides LIII," Collection Czech. Chem. Comm. **30**, 2592 (1965).

558—Clements, R. L., and St. J. Patterson: "Relationship between the 3,3′,5-triiodo-L-thyronine content of thyroid as determined by a TLC-method and the biological potency assayed by a rat anti-thiouracil Goitre method," Nature **207**, 1292 (1965).

559—Coenegracht, J., and T. Postmes: "Non-iodine containing compounds and (^{127}I)iodo-tyrosine-like substances studied by paper- and thin-layer chromatography," Clin. Chim. Acta **16**, 432 (1967).

560—Cohen, L. A., and L. Faber: "Cleavage of tyrosylpeptide bonds by electrolytic oxydation, in enzyme structure," Methods in Enzymology, Vol. XI, Academic Press, New York, 1967, 229.

561—Cole, M., and J. C. Fletcher: "The separation of iodobenzene-p-sulphonylamino acids (pipsylamino acids) by TLC," Biochem. J. **102**, 825 (1967).

562—Contractor, S. F., and J. Wragg: "Resolution of the optical isomers of dl-tryptophan, 5 hydroxy-dl-tryptophan and 6 hydroxy-dl-tryptophan by paper and thin layer chromatography," Nature **208**, 71 (1965).

563—Corbelli, A., F. Chillemi and P. G. Pieta: "Sintesi di peptidi della catena beta della emoglobina umana. Nota I. Sequenze 118-130, Nota II. Sequenze 131-138," Gazz. Chim. Ital. **97**, 514 (1967), **97**, 526 (1967).

564—Corbi, D., and L. Morselli: "Effect of ^{32}P on the urinary excretion of amino acids in the mouse. Method of determination and description of a technique for the qualitative and quantitative analysis," Giorn. Med. Milano **114**, 706 (1964).

565—Costopanagiotis, A. A., J. Preston and B. Weinstein: "Synthesis of the C-terminal tripeptide sequence (A$_{27}$-A$_{29}$) of glucogon," J. Org. Chem. **31**, 3398 (1966).

566—Costopanagiotis, A. A., J. Preston and B. Weinstein: "Amino acids and peptides V. VI," J. Org. Chem. **31**, 3398, 3400 (1966).

567—Criddle, W. J., G. J. Moody and J. D. Thomas: "Thin-Layer Electrophoresis," Lab. Pract. **15**, 653, 670 (1966).

568—Cunningham, B. A., P. D. Gottlieb, W. H. Konigsberg and G. M. Edelman: "The covalent structure of a human γG-immunoglobulin V," Biochemistry **7**, 1983 (1968).

569—Curtius, H. C., P. Anders, R. Zell, H. Sigel and H. Erlenmeyer: "Abbau des Ni^2-Polymixin-B-Komplexes durch H$_2$O$_2$," Helv. Chim. Acta **49**, 2256 (1966).

570—Cuzzino, M. T., and T. P. Lissi: "Identification of free amino acids in fruits as dinitrophenyl derivates," Farmaco [Pavia] Ediz. Prat. **20**, 488 (1965).

571—Czegledi-Janko, G.: "Determination of the degradation products of ethylene bis (dithiocarbamates) by TLC and some investigations of their composition in vitro," J. Chromatog. **31**, 89 (1967).

572—Davey, J. M., A. H. Laird and J. S. Morley: "Polypeptides. Part III," J. Chem. Soc. **1966**, 555.

573—Davis, V. E., J. L. Cashaw, J. A. Huff and H. Brown: "Identification of 5-OH-tryptophol as a serotonine metabolite in man," Proc. Soc. Exp. Biol. Med. **122**, 890 (1966).

574—Delacha, J. M., and A. U. Fontanive: "Quantitative N-terminal amino acid analysis by TLC," Experientia **21**, 351 (1965).

575—Derminot, J., and M. Tasdhomme: "Possible direct identification by chromatography of lanthionine and cysteic acid in the presence of cystine in wool hydrolysates," Bull. Inst. Textile France **117**, 247 (1965).

576—Determann, H.: "Untersuchungen über die Plastein-Reaktion VII," Liebig's Ann. **690**, 182 (1965).

577—Determann, H., J. Heuer, P. Pfaender and M. L. Reinartz: "Einfluss verschiedener N-Acylreste auf die Racemisierung bei Peptidsynthesen," Liebig's Ann. **694**, 190 (1966).

578—Deyl, Z., H. Molkova and J. Rosmus: "On the interaction between basic telopeptides and DNA," Experientia **23**, 723 (1967).

579—Deyl, Z., J. Rosmus and S. Bump: "Studies on the structure of collagen. II. The nature of the N- and C-terminals in enzyme-treated and untreated collagen," Biochim. Biophys. Acta **140**, 515 (1967).

580—Deyl, Z., L. Schinkmannova and J. Rosmus: "Chromatographic separation of 2,4-dinitro-5-aminophenyl derivatives of amino acids," J. Chromatog. **30**, 614 (1967).

581—Dezelee, P., and E. Bricas: "Nouvelle méthode de synthèse stéréospécifique de peptides de l'acide meso-a,a'-diamino-pimélique. I. Synthèse du dipeptide meso-diaminopimélyl-(L)-(D)-alanine," Bull. Soc. Chim. Biol. **49**, 1579 (1967).

582—Dietrich, C. P., A. V. Calucci and J. L. Strominger: "Biosynthesis of the peptidoglycan of bacterial cell walls V," J. Biol. Chem. **242**, 3218 (1967).

583—Dietrichs, H. H., and H. Funke: "Freie Aminosäuren im Phloem- und Frühjahrsblutungssaft der Rotbuche (Fagus sylvatica Linn.)," Holzforschung **21**, 102 (1967).

584—Dittmann, J.: "DC in der Klinik III: Zweidimensionale Trennung von Aminosäuren in biologischen Flüssigkeiten," Z. Klin. Chem. **4**, 8 (1966).

585—Dixon, H. B., and V. Moret: "Removal of the N-terminal residue of a protein after transamination," Biochem. J. **94**, 463 (1965).

586—Djurbabic, B., M. Vidakovic and D. Kolbah: "Effect of irradiation on the properties of pollen in Austrian and Scotch Pines," Experientia **23**, 236 (1967).

587—Doolittle, R. F., D. Schubert and S. A. Schwartz: "Amino acid sequence on arteriodactyl fibrinopeptides I," Arch. Biochem. Biophys. **118**, 456 (1967).

588—Dus, K., H. de Klerk, K. Sletten and R. G. Bartsch: "Chemical characterization of high potential iron proteins from chromatium and rhodipseudomonas gelatinosa," Biochim. Biophys. Acta **140**, 291 (1967).

589—Edelman, G. M., W. E. Gall, M. J. Waxdal and W. H. Konigsberg: "The covalent structure of a human γ G-immunoglobulin I," Biochemistry **7**, 1950 (1968).

590—Edman, P., and G. Begg: "A protein sequenator," Europ. J. Biochem. **1**, 80 (1967).

591—Eisler, K., J. Rudinger and F. Sorin: "Amino acids and peptides LXV," Collection Czech. Chem. Comm. **31**, 4562 (1966).

592—Eneroth, P., and G. Lindstedt: "TLC of betaines and other compounds related to carnitins," Anal. Biochem. **10**, 479 (1965).

593—Eriksson, B., and S. A. Eriksson: "Synthesis characterization of the L-cysteine-glutathione mixed disulfide," Acta Chem. Scand. **21**, 1304 (1967).

594—Erlenmeyer, H., H. Brintzinger, H. Sigel and H. C. Curtius: "Strukturspezifischer oxydativer Abbau von Peptid-Metall-Komplexen," Experientia **21**, 371 (1965).

595—Erlenmeyer, H., H. Sigel, H. C. Curtius and P. Anders: "Strukturspezifische Abbau-Reaktionen von Polypeptidmetall-Komplexen II. Abbau des Cu^{ii}-Polymyxin-B-Komplexes durch H_2O_2," Helv. Chim. Acta **49**, 19 (1966).

596—Esser, K.: "Ein dünnschicht-chromatographisches Verfahren zur quantitativen Bestimmung von Aminosäuren und Aminozuckern im Mikromasstab," J. Chromatog. **18**, 414 (1965).

597—Faircloth, M. A., A. D. Williams and W. H. Florsheim: "TLC method for the analysis of thyroidal iodoamino acids," Anal. Biochem. **12**, 437 (1965).

598—Fambrough, D. M., and J. Bonner: "On the similarity of plant and animal histones," Biochemistry **5**, 2563 (1966).

599—Fare, G., and D. C. Sammons: "TLC and Rf-values of amino acid-copper complexes in phenol/water," Experientia **22**, 668 (1966).

600—Farkasova, H., and J. Rudinger: "Amino acids and peptides LVII," Collection Czech. Chem. Comm. **30**, 3117 (1965).

601—Favino, A., D. Emrich and A. V. z. Muehlen: "Separation and quantitative determination of tri-iodo-thyronine- and thyroxine- [131]I in human plasma by TLC," Acta Endocrinol. **54**, 362 (1967).

602—Feder, J.: "Studies on the specificity of bacillus subtilis neutral protease with synthetic substrates," Biochemistry **6**, 2088 (1967).

603—Feltkamp, H., and H. Pfrommer: "Dünnschicht- und Säulenchromatographische Trennung einiger diastereomerer Dipeptide," J. Chromatog. **18**, 403 (1965).

604—Ferguson, D. R.: "Separation of mammalian neurohypophysial hormones by TLC," J. Endocrinol. **32**, 119 (1965).

605—Fernlund, P., and L. Josefsson: "Chromactivating hormones of Pandalus Borealis," Biochim. Biophys. Acta **158**, 262 (1968).

606—Flierl, C., H. Sigel and H. Erlenmeyer: "Strukturspezifischer Abbau von Polypeptid-Metall-Komplexen V," Experientia **22**, 784 (1966).

607—Fölsch, G.: "Synthesis of phosphopeptides V," Acta Chem. Scand. **20**, 459 (1966).

608—Fontana, A., F. Machiori, R. Rocchi and P. Pagetta: "Protezione del gruppo amminico mediante residui solfenici nella sintesis di peptidi I, II," Gazz. Chim. Ital. **96**, 1301, 1313 (1966).

609—Fosker, A. P., and H. D. Law: "Oxytoxin and 4-glycine oxytoxin," J. Chem. Soc. **1965**, 4922.

610—Fosker, A. P., and H. D. Law: "L-Glutamyl-γ-aminobutyric acid and related compounds," J. Chem. Soc. **1965**, 7305.

611—Franek, F., B. Keil, J. Novotny and F. Sorm: "Amino acid sequence around disulfide bridges of pig immunoglobin λ-chains," Europ. J. Biochem. **3**, 422 (1968).

612—Franze de Fernandez, M. T., A. C. Paladini and A. E. Delius: "Isolation and identification of Pepsitensin," Biochem. J. **97**, 540 (1965).

613—Frei, R. W., and M. M. Frodyma: "A review on recent developments in the analysis of amino acids and other systems by PC and TLC," Chem. Rundschau [Solothurn] **12**, 26 (1966).

614—Frentz, R.: "TLC of urinary amino acids," Ann. Biol. Clin. [Paris] **23**, 1145 (1965).

615—Frey, H.: "Quantitative determination of thyroid gland products," Scand. J. Clin. Lab. Invest. **16**, 470 (1964).

616—Fridkin, M., A. Patchornik and E. Katscholski: "Synthesis of cyclic peptides," J. Am. Chem. Soc. **87**, 4646 (1965).

617—Gall, W. E., B. A. Cunningham, M. J. Waxdal, W. H. Konigsberg and G. M. Edelman: "The covalent structure of a human γG-immunoglobulin IV," Biochemistry **7**, 1973 (1968).

618—Gallop, P., O. O. Blumenfeld, E. Henson and A. L. Schneider: "Isolation and identification of a-amino-aldehydes in collagen," Biochemistry **7**, 2409 (1968).

619—Gansser, C.: "Acides dicarboxyliques a,a'-diaminés et composés apparentés," Bull. Soc. Chim. France **1966**, 1713.

620—Gebert, U., T. Wieland and H. Boehringer: "Die Konstruktion von Amanin und Phallisin," Ann. Chem. **705**, 227 (1967).

621—Gerber, G. B., and J. Remy-Defraigue: "A micro method for the quantitative determination of beta-aminoisobutyric acid," Anal. Biochem. **11**, 386 (1965).

622—Gerwing, J., B. Mitchell and D. van Alstyne: "Studies on the active region of botulinus toxins. II. Isolation and amino acid sequence of the cysteine-containing tryptic peptides in botulinus toxins types A, B and E," Biochim. Biophys. Acta **140**, 363 (1967).

623—Goedde, H. W.: "Beta-aminobutyric acid: A TLC method for the quantitative estimation in human urine," Clin. Chim. Acta **11**, 485 (1965).

624—Ghuysen, J. M., D. J. Tipper, C. H. Birge and J. L. Strominger: "Structure of the cell wall of staphylococcus aureus Strain Copenhagen," Biochemistry **4**, 2245 (1965).

625—Ghuysen, J. M., E. Bricas, M. Leiyh-Bouille, M. Lache and G. D. Shockman: "The peptide N^a-(L-alanyl-D-isoglutamyl)-N_e-(D-isoasparaginyl)-L-lysyl-D-alanine and the disaccharide N-acetylglucosaminyl-β-1,4-N-acetyl-muramic acid in cell wall peptidoglycan of Streptococcus faecalis Strain ATCC 9790," Biochemistry **6**, 2607 (1967).

626—Goodman, M., and W. J. McGahren: "Mechanistic studies of peptide oxazolone racemization," Tetrahedron **23**, 2031 (1967).

627—Gorchein, A.: "Studies on the structure of an ornithine-containing lipid from non-sulphur purple bacteria," Biochim. Biophys. Acta **152**, 358 (1968).

628—Gräsbeck, R., K. Simons and I. Sinkkonen: "Isolation of intrinsic factor and its probable degradation product as their vitamin B 12 complexes, from human gastric juice," Biochim. Biophys. Acta **127**, 47 (1966).

629—Gregory, H., A. H. Laird, J. S. Morley and J. M. Smith: "Polypeptides. Part VI. Variations of the terminal amide position in the C-terminal tetrapeptide amide sequence of the gastrins. Part VII. Variations of the phenylalanine position in the C-terminal tetrapeptide amide sequence of the gastrins," J. Chem. Soc. **1968**, 522, 531.

630—Gregory, H., J. S. Morley, J. M. Smith and M. J. Smithers: "Polypeptides VIII. Variations of the aspartyl position in the C-terminal tetrapeptide amide sequence of the gastrins. Polypeptide IX. Variations of the methyonyl position in the C-terminal tetrapeptide amide sequence of the gastrins," J. Chem. Soc. **1968**, 715, 726.

631—Gros, C.: "Microanalyse des acides aminés et peptides. Réactivité des acides aminés et peptides avec le chlorure de dansyle et séparation des dérivés dansylés formés," Bull. Soc. Chim. France **1967**, 3952.

632—Grupe, R., and H. Niedrich: "Hydrazinverbindungen als Hetero-bestandteile in Peptiden VII," Chem. Ber. **99,** 3914 (1966).

633—Guinand, M., and G. Michel: "Structure d'un peptidolipide isolé de Nocardia Asteroides, la peptidolipine NA," Biochim. Biophys. Acta **125,** 75 (1966).

634—Guinand, M., M. J. Vacheron, G. Michel, B. C. Das and E. Lederer: "Détermination de séquences d'acides aminés des oligopeptides par la spectrométrie de masse V," Tetrahedron Suppl. **7,** 271 (1966).

635—Guttmann, S.: "Synthèse du glutathion et de l'oxytocine à l'aide d'un nouveau groupe protecteur de la fonction thiol," Helv. Chim. Acta **49,** 83 (1966).

636—Habermann, E., and J. Jentsch: "Sequenzanalyse des Mellitins aus den tryptischen und peptischen Spaltstücken," Z. Physiol. Chem. **348,** 37 (1967).

637—Habermann, E., and K. G. Reiz: "Zur Biochemie der Bienengift-peptide Melitin und Apamin," Biochem. Z. **343,** 192 (1965).

638—Habermann, E., and J. Helbig: "Untersuchung zur Struktur des Rindserum-Kininogens unter Verwendung von Bromcyan und Car-boxypeptidase B," Naunyn-Schmiedebergs Arch. Exp. Pathol. Pharmakol. **258,** 160 (1967).

639—Haleysovsky, V., B. Mesrob, V. Tomasek and F. Sorm: "Peptides from peptic digest of bovine DIP-trypsin and determination of disulfide bonds," Collection Czech. Chem. Comm. **33,** 441 (1968).

640—Hamberg, U.: "Formulation of active peptide structure by quantitative end group determination and bioassay," Acta. Chem. Scand. **20,** 274 (1966).

641—Handford, B. O., J. H. Jones, G. T. Young and T. F. Johnson: "Amino acids and peptides XXIV," J. Chem. Soc. **1965,** 6814.

642—Haschenmeyer, R. H., M. A. Cynkin, L. C. Han and M. Trindle: "Isolation and amino acid sequences of glycopeptides obtained from bovine fibrinogen," Biochemistry **5,** 3443 (1966).

643—Hashmi, M. H., A. S. Adil, N. A. Chughtai, F. R. Chughtai and M. A. Shahid: "Identification of twenty amino acids by circular TLC," Mikrochim. Acta **1968,** 291.

644—Hassall, C. H., T. G. Martin, J. A. Schofield and J. O. Thomas: "The synthesis, by twinning, of cyclodepsipeptides related to serratomo-lide," J. Chem. Soc. **1967,** 997.

645—Hatch, F. T., Y. Aso, L. M. Hagopian and J. J. Rubenstein: "Biosynthesis of lipoprotein by rat intestinal mucosa," J. Biol. Chem. **241,** 1655 (1966).

646—Hauton, J. C., B. Laurent, A. Gerolami Santandrea, C. Greusard, H. Lafont, N. Tessier and H. Sarles: "Etude préliminaire d'une fraction peptidique extraite avec les lipides neutres du plasma," Clin. Chim. Acta **17,** 171 (1967).

647—Haux, P., H. Sawerthal and E. Habermann: "Sequenzanalyse des Bienengift-Neurotoxins (Apamin) aus seinen tryptischen und chymotryptischen Spaltstücken," Z. Physiol. Chem. **348,** 737 (1967).

648—Heathcote, J. G., and K. Jones: "Rapid resolution of complex mixtures of the naturally occurring amino acids by two-dimensional TLC on cellulose powder," Biochem. J. **97,** 15 P (1965).

649—Heider, J. G., and J. R. Brouk: "Rapid separation of thyroxine and some of its analogues by TLC," Biochim. Biophys. Acta **95,** 353 (1965).

650—Heigenoort, J. van, et al: "Détermination de séquences d'acides aminés dans des oligopeptides par la spectrométrie de masse. IX," Tetrahedron **23,** 3403 (1967).

651—Heller, W.: "Tonmineralien bituminöser Schiefer als natürliche Systeme der Verteilungschromatographie," Erdöl Kohle **19,** 557 (1966).

652—Herberhold, C., and O. A. Neumüller: "DC-Unters. des Lymphbahninhalts menschlicher Schilddrüsen: Trennung von Jodaminosäuren aus autopischem Material," Klin. Wochschr. **43,** 717 (1965).

653—Heyns, K., and Hauber: "DC-Trennung und quantitative Ultramikrobestimmung von Aminosäuren als DNP-Aminosäure- (^{14}C) methylester nach der Szintillationsmethode," Z. Physiol. Chem. **348,** 357 (1967).

654—Hill, R. D., and R. R. Laing: "Specific reaction of dansylchloride with one lysine residue in rennin," Biochim. Biophys. Acta **132,** 188 (1967).

655—Hirano, K.: "Determination of free amino acids in the urine of mentally retarded children by TLC," Osaka Shiritsu Daigaku Kaseigakubu Kiyo **12,** 115 (1964).

656—Hisatsune, K., S. J. de Courcy and S. Mudd: "The immunologically active cell wall peptide polymer of staphylococcus aureus," Biochemistry **6,** 595 (1967).

657—Hiskey, R. G., and J. B. Adams, Jr.: "Sulfur-containing polypeptides I," J. Am. Chem. Soc. **87,** 3969 (1965).

658—Hiskey, R. G., and J. B. Adams, Jr., "Sulfur-containing polypeptides IV," J. Org. Chem. **31,** 2178 (1966).

659—Hiskey, R. G., T. Mizoguchi and H. Igeta: "Sulfur-containing polypeptides II," J. Org. Chem. **31,** 1188 (1966).

660—Hochstrasser, K., and E. Werle: "Ueber kininliefernde Peptide aus pepsinverdauten Rinderplasmaproteinen," Z. Physiol. Chem. **348,** 177 (1967).

661—Hoenders, H. J., and H. Bloemendal: "The N-terminus of the lens protein a-crystallin," Biochim. Biophys. Acta 147, 183 (1967).

662—Holbrook, J. J., et al.: "The amino acid sequence around the essential SH-group of pig heart LDH, isoenzyme I," Europ. J. Biochem. 1, 476 (1967).

663—Holcomb, G. N., S. A. James and D. N. Ward: "A critical evaluation of the selective tritiation method of determining C-terminal amino acids and its application to luteinizing hormone," Biochemistry 7, 1291 (1968).

664—Holloway, C. T., R. P. Bond, I. G. Knight and R. B. Beechey: "The alleged presence and role of monoiodonistidine in mitochondrial oxidative phosphorylation," Biochemistry 6, 19 (1967).

665—Holm, H., L. O. Froholm and S. Laland: "Isolation of peptide conjugate with the sequence Phe-Pro-Val-Orn from a cell-free system producing Gramicidin S," Biochim. Biophys. Acta 115, 361 (1966).

666—Hori, K., and M. J. Cormier: "Studies on the bioluminescence of Renilla Reniformis VI. Some chemical properties and the tentative partial structure of luciferin," Biochim. Biophys. Acta 130, 420 (1966).

667—Hörmann, H., K. T. Joseph and Mv. Wilm: "Ueber Aminoendgruppen von Kollagen III," Z. Physiol. Chem. 341, 284 (1965).

668—Horstmann, H. J., and L. Gürtler: "Vergleich der Peptidmuster einiger mit Trypsin hydrolysierter ribosomaler Proteine der Bäckerhefe," Z. Physiol. Chem. 349, 410 (1968).

669—Horton, D., A. Tanimura and M. L. Wolfrom: "Two-dimensional TLC of amino acids on microcrystalline cellulose," J. Chromatog. 23, 309 (1966).

670—Howlett, M. R., and G. B. Selzer: "The identification of colistin and polymyxin B by TLC," J. Chromatog. 30, 630 (1967).

671—Huguenin, R. L., and R. A. Boissonas: "Synthèse de la désamino[1]-Arg[8]-vasopressine et de la désamino[1]-Phe[2]-Arg[8]-vasopressine," Helv. Chim. Acta 49, 695 (1966).

672—Huguenin, R. L.: "Synthèse de la désamino[1]-Orn[8]-vasopressine, de la désamino[1]-Phe[2]-Orn[8]-vasopressine, de la désamino[1]-Ile[3]-Orn[8]-vasopressine et de la désamino[1]-Phe[2]-Orn[8]-vasopressine," Helv. Chim. Acta 49, 711 (1966).

673—Igloy, M., and A. Mizsei: "Separation of polymyxin B, D, E and M by TLC," J. Chromatog. 28, 456 (1967).

674—Ikehawa, N., O. Hoshino, R. Watanuki, H. Orino, T. Fujita and M. Yoshikawa: "Gas-chromatographic separation of DNP-amino acids," Anal. Biochem. 17, 16 (1966).

675—Inouye, M., and A. Tsugita: "The amino acid sequence of T 4 bacteriophage Lysozyme," J. Mol. Biol. 22, 193 (1966).

676—Inouye, K., I. M. Voynick, G. R. Delpierre and J. S. Fruton: "New synthetic substrates for pepsin," Biochemistry 5, 2473 (1966).

677—Inouye, K., and J. S. Fruton: "Studies on the specificity of pepsin," Biochemistry 6, 1765 (1967).

678—Iwasaki, I.: "Clinical application of TLC 3: Analysis of free amino acids in ascites and hydrothorax fluids," Med. Biol. [Tokyo] 71, 17 (1965).

679—Iwasaki, I.: "Clinical application of TLC 1: Analysis of free amino acids in blood plasma," Med. Biol. [Tokyo] 70, 24 (1965).

680—Iwasaki, I., S. Arimori and E. Watanabe: "Clinical application of thin-layer chromatography. 2: Analysis of free amino acids in erythrocytes," Med. Biol. [Tokyo] 70, 205 (1966).

681—Jakubke, H. D., and A. Voigt: "Ueber aktivierte Ester VII + VIII," Chem. Ber. 99, 2419, 2944 (1966).

682—Janecke, H., and K. Schaffnit: "Ueber Wollfett," Pharmazie 22, 93 (1967).

683—Jaquenoud, P. A.: "Synthèse de la Gly^8-oxytocine, de l'Ala^8-oxytocine et de la But^8-oxytocine," Helv. Chim. Acta 48, 1899 (1965).

684—Jarvis, D., and J. L. Strominger: "Structure of the cell wall of staphylococcus aureus VIII. Structure and chemical synthesis of the basic peptide released by the myxabacterium enzyme," Biochemistry 6, 2591 (1967).

685—Jeppsson, J. O., and J. O. Sjöquist: "TLC of PTH amino acids," Anal. Biochem. 18, 264 (1967).

686—Jolles, G., and P. Jolles: "Isolement et caractérisation d'un facteur anthistaminique présent dans certains lots de lysozyme," Bull. Soc. Chim. France 1965, 3300.

687—Jolles, G., and P. Jolles: "Human tear and human milk lysozymes," Biochemistry 6, 411 (1967).

688—Jolles, G., G. Poiget, J. Robert, B. Terlain and J. P. Thomas: "Pristinamycin: Synthesis of the linear heptapeptide and oligopeptides corresponding to the I$_A$ constituent of prystynamycin," Bull. Soc. Chim. France (8) 1965, 2252.

689—Jones, K., and J. G. Heathcote: "The rapid resolution of naturally occurring amino acids by TLC," J. Chromatog. 24, 106 (1966).

690—Josselin, J., J. Gombert, R. Masseyeff: "N-terminale Aminosäuren des Makroimmunoglobulins," Compt. Rend. 260, 3519 (1965).

691—Jost, K.: "Amino acids and peptides LXI," Collection Czech. Chem. Comm. 31, 2784 (1966).

692—Jürgens, P., H. W. Bansi and G. Müller: "Der Einfluss anabolier Hormone auf den Eiweiss- und Energie-Stoffwechsel des Menschen," Klin. Wochschr. 44, 165 (1966).

693—Käser, H., and E. Gugler: "DC-Bestimmung der $_\epsilon$-Aminocapronsäure in Serum und Urin," Z. Klin. Chem. **3**, 33 (1965).

694—Kahnt, F. W., B. Riniker, I. MacIntyre and R. Neher: "Thyrocalcitonin I. Isolierung und Charakterisierung wirksamer Peptide aus Schweineschilddrüsen," Helv. Chim. Acta **51**, 214 (1968).

695—Kanazawa, T., K. Kanazawa and Y. Morimura: "New arginine-containing peptides isolated from Chlorella cells," Plant Cell Physiol. [Tokyo] **6**, 831 (1965).

696—Karmarkar, M. G., and J. B. Stanbury: "The role of pyridoxal and manganous ions in the synthesis of iodothyronines," Biochim. Biophys. Acta **141**, 483 (1967).

697—Kasafirek, E., K. Jost, J. Rudinger and F. Sorin: "Amino acids and peptides LIV," Collection Czech. Chem. Comm. **30**, 2600 (1965).

698—Kasafirek, E., V. Rabek, J. Rudinger and F. Sorin: "Amino acids and peptides LXVI," Collection Czech. Chem. Comm. **31**, 4581 (1966).

699—Kasuya, M., and H. Takashina: "Tryptic digestion of Myosin A labeled with DNS-Cl," J. Biochem. [Tokyo] **60**, 459 (1966).

700—Kataura, A., and K. Kataura: "The comparison of free and bound amino acids between dry and wet types of cerumen," Tohoku J. Exp. Med. **91**, 215 (1967).

701—Katrukha, G. S., Z. P. Trifonova and A. B. Silaev: "Amino acid analysis of proteins and peptides by the method of thin-layer electrophoresis combined with chromatography," Khim. Prir. Soedin. **2**, 348 (1966).

702—Katsoyannis, P. G., A. Tometsko, C. Zalut, S. Johnson and A. C. Trakatellis: "Studies on the synthesis of insulin from natural and synthetic A and B chains. I. Splitting of insulin and isolation of the s-sulfonated derivatives of the A and B chains," Biochemistry **6**, 2635 (1967).

703—Katsoyannis, P. G., A. C. Trakatellis, S. Johnson, C. Zalut and G. Schwartz: "Studies on the synthesis of insulin from natural and synthetic A and B chains. II. Isolation of insulin from recombination mixtures of natural A and B chains," Biochemistry **6**, 2642 (1967).

704—Katsoyannis, P. G., A. C. Trakatellis, C. Zalut, S. Johnson, A. Tometsko, G. Schwartz and J. Ginos: "Studies on the synthesis of insulin from natural and synthetic A and B chains. III. Synthetic insulin," Biochemistry **6**, 2656 (1967).

705—Katsoyannis, P. G., A. M. Tometsko and C. Zalut: "Insulin peptides XVII. The synthesis of the A chain of human (porcine) insulin and its isolation as the S-sulfonated derivative," J. Am. Chem. Soc. **89**, 4505 (1967).

706—Katz, W., M. Matsuhashi, C. P. Dietrich and J. L. Strominger:

"Biosynthesis of peptidoglycans of bacterial cell walls IV," J. Biol. Chem. **242**, 3207 (1967).

707—Katz, W., and J. L. Strominger: "Structure of the cell wall of micrococcus lysodeikicus II," Biochemistry **6**, 930 (1967).

708—Katz, S., and A. Lewis: "Apparatus for HVTLE: application to amino acid analysis," Anal. Biochem. **17**, 300 (1966).

709—Katze, J., S. Sridhara and I. Zabin: "An amino-terminal peptide of beta-galactosidase," J. Biol. Chem. **241**, 5341 (1966).

710—Keenan, T. W., and R. C. Lindsay: "Rapid resolution of methylmethionine sulfonium salts and homoserine by TLC," J. Chromatog. **30**, 251 (1967).

711—Kessler, W., and B. Iselin: "Selektive Spaltung substituierter Phenylsulfenyl-Schutzgruppen bei Peptidsynthesen," Helv. Chim. Acta **49**, 1330 (1966).

712—Khosla, M. C., R. R. Smeby and F. M. Bumpus: "Solid-phase peptide synthesis of (L-alanine³-L-isoleucine⁵)-angiotensin II," Biochemistry **6**, 754 (1967).

713—King, T. P.: "Selective chemical modifications of arginyl residues," Biochemistry **5**, 3454 (1966).

714—Kinoshita, M., and H. Klostermeyer: "Synthese von Heptapeptid-Derivaten mit der Insulinsequenz B 14-20," Liebig's. Ann. **696**, 226 (1966).

715—Klostermann, H. J., G. L. Lamoureux and J. L. Parsons: "Isolation, characterization and synthesis of linatine, a vitamin B₆ antagonist from flaxseed (linum usitatissimum)," Biochemistry **6**, 170 (1967).

716—Köchel, F., and P. Frank: "Entwicklung und Prüfung einer bilanzierten Aminosäure-Infusionslösung," Krankenhaus-Apotheke **15**, 17 (1965).

717—Koning, P. J. de, P. J. van Rooijen: "Amino acid composition of aS_1-casein D," Nature **213**, 1028 (1967).

718—Kredich, N. M., and G. M. Tomkins: "Enzymic synthesis of L-cysteine in Escherichia coli and Salmonella typhimurium," J. Biol. Chem. **241**, 4955 (1966).

719—Krenitsky, T. A., and J. S. Fruton: "An aminoacyltransferase preparation from beef liver. J. Biol. Chem. **241**, 3347 (1966).

720—Kress, L. F., P. J. Peanasky and H. M. Klitgaard: "Purification, properties and specifity of hog thyroid proteinase," Biochim. Biophys. Acta **113**, 375 (1966).

721—Kycia, J. H., M. Elzinga, N. Alonzo and C. H. Hirs: "Primary structure of bovine carboxypeptidase BI," Arch. Biochem. Biophys. **123**, 336 (1968).

722—Lai, C. Y., P. Hoffee and B. L. Horecker: "Mechanism of action of Aldolases XII," Arch. Biochem. Biophys. **112**, 567 (1965).

723—Lai, C. Y., C. Chen and O. Tsolas: "Isolation and sequence analysis of a peptide from the active site of transaldolase," Arch. Biochem. Biophys. **121**, 790 (1967).

724—Lambiotte, M.: "Thin-layer autoradiographic electrophoresis of tritium-labeled compounds in buffered photographic gelatin," Atomlight No. 45, 10 (1965).

725—Langner, J.: "Eine dünnschichtchromatographische Fingerprint-methode für Peptide unter Verwendung von DNS-Cl," Z. Physiol. Chem. **347**, 275 (1966).

726—Lapidot, Y., N. deGroot, M. Weiss, R. Peled and Y. Wolman: "The synthesis of glycyl-L-phenylalanyl-S-RNA," Biochim. Biophys. Acta **138**, 241 (1967).

727—Laufer, D. A., and E. R. Blout: "A new stepwise synthesis of an octapeptide corresponding to a sequence around the 'reactive' serine of chymotrypsin," J. Am. Chem. Soc. **89**, 1246 (1967).

728—Laves, W., and A. Winkler: "DC-Untersuchungen an vitalen und postmortalen Blutseren," Dtsch. Z. Ges. Gerichtl. Med. **57**, 424 (1966); ref. Chem. Zbl. **138**, 2133, p. 240 (1967).

729—Lees, C. W., and R. W. Hartley: "Studies on bacillus subtilis ribonuclease III," Biochemistry **5**, 3951 (1966).

730—Lehrer, S. S., and G. D. Fasman: "Ultraviolet irradiation effects in poly-L-tyrosine and model compounds. Identification of bityrosine as photoproduct," Biochemistry **6**, 757 (1967).

731—Leutgeb, W., and U. Schwarz: "Abbau des Mureins als erster Schritt beim Wachstum des Sacculus," Z. Naturforsch. **22**, 545 (1967).

732—Ligny, C. L. de, and A. G. Remijnse: "Efficiency of chromatographic procedures IV," Rec. Trav. Chim. **86**, 421 (1967).

733—Lin, Y. T., K. T. Wang and I. S. Wang: "Polyamide layer chromatography: Identification of amino acids in angelica acutiloba Kitagawa via DNP-Derivatives," J. Chinese Chem. Soc. [Taiwan] **13**, 19 (1966).

734—Labouesse, J., and M. Gervais: "Preparation of defined εN-acetylated trypsin," Europ. J. Biochem. **2**, 215 (1967).

735—Lindberg, U.: "Molecular weight and amino acid composition of deoxyribonuclease I," Biochemistry **6**, 335 (1967).

736—Llosa, P. de la, C. Harmier, P. A. de la Llosa and M. Jutisz: "Composition of hypophysal FSM of sheep and investigation of the terminal amino acids," Bull. Soc. Chim. Biol. **47**, 1073 (1965).

737—Lorentz, K., and A. Hartmann: "TLC of carbamoyl compounds," J. Chromatog. **30**, 250 (1967).

738—Majerus, P. W.: "Acyl carrier proteins," J. Biol. Chem. **242**, 2325 (1967).

739—Manning, M., and V. du Vigneaud: "6-Hemi-D-cystine-oxytocin," J. Am. Chem. Soc. **87**, 3978 (1965).
740—Marchiori, F., R. Rocchi, G. Vidali, A. Tamburro and E. Scoffone: "Synthesis of peptides analogues to the N-terminal eicosapeptide sequence of ribonuclease A," J. Chem. Soc. **1967**, 81, 86, 89.
741—Marcucci, F., and E. Mussini: "Separation of proline and hydroxyproline derivatives by TLC," J. Chromatog. **18**, 431 (1965).
742—Mardarowocz, C.: "Isolierung und Charakterisierung des Murein-Sacculus von Brucella," Z. Naturforsch. **21**, 1006 (1966).
743—Mardashev, S. P., L. S. Semina, F. N. Prozorovskij and A. M. Sokhina: Biochimija **32**, 761 (1967).
744—Mardashev, C. R., M. Z. Zaleskij, S. I. Pechtereva and L. A. Semina: Biochimija **32**, 655.
745—Markwardt, F., H. P. Klöcking and M. Richter: "Die Bestimmung synthetischer Antifibrnolytica in Körperflüssigkeiten und Organen," Pharmazie **22**, 83 (1967).
746—Markwardt, F., and P. Walsmann: "Reindarstellung und Analyse des Thrombininhibitors Hirudin," Z. Physiol. Chem. **348**, 1881 (1967).
747—Markwardt, F., W. Barthel, E. Glusa, A. Hoffmann and P. Walsmann: "Ueber eine aminfreisetzende Komponente des Crotalus terrificus-Giftes," Biochem. Z. **346**, 351 (1966).
748—Marszalek, J.: "Recherches sur les phénylthiohydantoines et leur emploi dans la détermination de séquences peptidiques," Dissertation, Universität Genève (1965).
749—Martin, H., and L. Nowicki: "Hämoglobinopathien, Bedeutung und Nachweismethoden," Glas.-Instr.-Tech. **1967**, 320.
750—Martinez-Carrion, M., and D. Tiemeier: "Mitochondrial glutamate-aspartate transaminase I. Structural comparison with the supernatant isozyme," Biochemistry **6**, 1715 (1967).
751—Marzona, M., and D. di Modica: "Determination of lanthionine by TLC," J. Chromatog. **32**, 755 (1968).
752—Maskaleris, M. L., E. S. Sevendal and A. C. Kibrick: "Carbobenzoxy derivatives of amino acids and peptides: Instant TLC as hydrobromides," J. Chromatog. **23**, 403 (1966).
753—Massaglia, A., and U. Rosa: "Separation of the ^{131}I-labelled S-sulfonated A and B insulin chains by TLC," J. Label. Comp. **1**, 141 (1965).
754—Mathur, K. B., H. Klostermeyer and H. Zahn: "Synthese von Dekapeptid-Derivaten der Insulinsequenz B_{21-30}," Z. Physiol. Chem. **346**, 60 (1966).
755—Matsuhashi, M., C. P. Dietrich and J. L. Strominger: "Biosynthesis of the peptidoglycan of bacterial cell walls III," J. Biol. Chem. **242**, 3191 (1967).

756—Mattei, P. di: "Occurrence of bradykinin in human pulmonary carcinoma," Pharmacol. **16,** 909 (1967).

757—Meienhofer, J.: "Synthese von Gly⁷-cyclo-Kallidin," Liebig's Ann. **691,** 218 (1966).

758—Melani, F., R. Guazelli, F. Salti and U. Bigozza: "Separazione e dosaggio delle iodotironine e delle iodotirosine mediante crom. su strato sottile," Giorn. Biochim. [Ital.] **13,** 376 (1964/65).

759—Mellon, J. M., and A. G. Stiven: "Rapid method for the detection of plasma phenylalanine," J. Med. Lab. Tech. **23,** 204 (1966).

760—Meloun, B., I. Fric and F. Sorm: "Nitration of tyrosine residues in the pancreatic trypsin inhibitor with tetranitromethane," Europ. J. Biochem. **4,** 112 (1968).

761—Meloun, B., L. Moravek and F. Sorm: "On proteins CVII. Peptides isolated from B- and C-chain of S-carboxymethyl-DIP-a-Chymotrypsin," Collection Czech. Chem. Comm. **32,** 1947 (1967).

762—Mesrob, B., and V. Holeysovsky: "Differentiation of dimethylamino-aphthalenesulphonic derivatives (DNS) of valine, leucine and isoleucine by chromatography on a thin layer of silica gel," J. Chromatog. **21,** 135 (1966).

763—Mesrob, B., and V. Holeysovsky: "Microdetermination of C-terminal groups of peptides and proof of amides in terminal groups," Collection Czech. Chem. Comm. **32,** 1976 (1967).

764—Meyer, J., K. Droll and V. Klingmüller: "Ein Suchtest zur Erkennung der Phenylketonurie und eine semiquantitative Bestimmung des Phenylalanins im Blutplasma mit der Dünnschichtchromatographie," Clin. Chim. Acta **18,** 69 (1967).

765—Meynel, G., J. A. Berger, R. N. Arnaud and P. Banquet: "Etudes du comportement de quelques composés iodés thyroidiens sur films chrom. souples," Compt. Rend. **260,** 3065 (1965).

766—Miersch, J.: "Nachweis und Isolierung von Canalin," Naturwissenschaften **54,** 169 (1967).

767—Miersch, J., and H. Reinbothe: "Chromatographic separation of amino acids and guanidine compounds from fruit-body of higher fungi," Flora [Jena] **156,** 543 (1966).

768—Mikes, O., H. G. Müller, V. Holeysovsky, V. Tomasek and F. Sorin: "Peptides isolated from chymotriptic digest of S-sulfo-DIP-trypsin," Collection Czech. Chem. Comm. **32,** 620 (1967).

769—Millar, K. R.: "Separation of methionine and selenomethionine by TLC," J. Chromatog. **21,** 344 (1966).

770—Millstein, S., and D. W. Thomas: "Separation of thyroid hormones by TLC and their quantitative measurement," J. Lab. Clin. Med. **67,** 496 (1966).

771—Mirelman, D., and N. Sharon: "Isolation and study of the chemical structure of low molecular weight glycopeptides from micrococcus lysodeikticus cell walls," J. Biol. Chem. **242**, 3414 (1967). TLC on silica with 5 solvent systems for the identification of free and DNP amino acids.

772—Miyaji, T., I. Iuchi, K. Yamamoto, Y. Ohba and S. Shibata: "Amino acid substitution of hemoglobin UBE 2 ($a_2^{68'Asp}\beta_2$): an example of successful application of partial hydrolysis of peptide with 5% acetic acid," Clin. Chim. Acta **16**, 347 (1967).

773—Moravek, L., I. Kluh, J. M. Junge, B. Meloun and F. Sorin: "Structure of peptides isolated from chymotryptic digest of bovine S-carboxymethyl-chymotrypsinagen A," Collection Czech. Chem. Comm. **31**, 1142 (1966).

774—Morley, J. S.: "Synthesis of human gastrin," J. Chem. Soc. **1967**, 2410. Synthesis of a 13-peptide. TLC with 8 solvent systems acc. to Morley et al. J. Chem. Soc. (L) **1966**, 555.

775—Moroz, C., A. de Vries and M. Sela: "Isolation and characterization of a neutrotoxin from vipera Palestinae venom," Biochim. Biophys. Acta **124**, 136 (1966).

776—Moser, J. G.: "Reinigung und Charakterisierung einer Proteinase der Larve von Calliphora erythrocephala Meigen," Biochem. Z. **344**, 337 (1966).

777—Most, C. F., Jr., and H. B. Milne: "Quantitative TLC of benzyloxycarbonyl dipeptide phenylhydrazides," J. Chromatog. **34**, 551 (1968).

778—Munier, R. L., and C. Thommegay: "Chromato-électrophorèse des aminoacides en c.m. de poudre de cellulose. IV. Procédé analytique à haut pouvoir séparateur pour les aminoacides de faible mobilité (alanine, asparagine, glutamine, glycocolle, méthionine sulfone, méthionine sulfoxyde, proline, sérine, thréonine) et des amino-diacides (acide aspartique, acide cystéique, acide glutaminique, S-carboxyméthylcystéine) V. Procédé à haut pouvoir séparateur permettant l'identification de tous les aminoacides présents dans les hydrolysats chimiques ou enzymatiques de protéines modifiées ou non chimiquement," Bull. Soc. Chim. France **1967**, IV:3171, V:3176.

779—Munier, R. L., C. Thommegay and G. Sarrazin: "Méthodes d'analyse systématique des mélanges d'amino-acides par chromatographie et chromato-électrophorèse en c.m. de poudre de cellulose," Bull. Soc. Chim. France **1967**, 3971.

780—Munier, R. L., C. Thommegay and A. M. Drapier: "Separation of dansyl-amino acids by TLC on cellulose powder," Chromatographia **1**, 95 (1968).

781—Munoz, E., J. M. Ghuysen, M. Leyh-Bouille, J. F. Petit, H. Heymann, E. Bricas and P. Lefrancier: "The peptide subunit Na-(L-alanyl-D-isoglutaminyl)-L-lysyl-alanine in cell wall peptidoglycans of staphylococcus areus strain Copenhagen, micrococcus roseus R 27, and streptococcus pyogenes group A, type 14," Biochemistry 5, 3748 (1966).

782—Murray, M., and G. F. Smith: "Infrared spectra of microgram amounts of amino acid phenylthiohydantoins eluted from thin layers of silica gel," Anal. Chem. 40, 440 (1968).

783—Musha, S., and H. Ochi: "Separation of amino acids by TLC," Japan Analyst 14, 202 (1965).

784—Musha, S., and H. Ochi: "Sodium carboxymethylcellulose in TLC," Bunseki Kagaku 14, 202 (1965).

785—Musha, S., and H. Ochi: "Separation of aromatic amino acids by means of activated carbon TL-plates and application of migration electrography to its detection," Bunseki Kagaku 14, 731 (1965); ref. Chem. Abstr. 63, 15205G (1965).

786—Nagasawa, S., Y. Mizushima, T. Sato, S. Iwanaga and T. Suzuki: "Studies on the chemical nature of bovine bradyknogen," J. Biochem. [Tokyo] 60, 643 (1966).

787—Nast, H. P., and H. Fasold: "Rasche Fingerprint-Entwicklung auf Kieselgel-Dünnschichtplatten," J. Chromatog. 27, 499 (1967).

788—Nedkov, P., and N. Genov: "Extending the range of application of the Edman method. Application to short peptides in small amounts," Biochim. Biophys. Acta 127, 541 (1966).

789—Nesvabda, H., H. Bachmayer and H. Michl: "Synthese eines im Gift von Bombina variegata vorkommenden Hexapeptid-diamids," Monatsh. Chem. 96, 1125 (1965).

790—Niall, H. D., and P. Edmann: "Two structurally distinct classes of kappa-chains in human immunoglobulins," Nature 216, 262 (1967).

791—Niedrich, H.: "Hydrazinverbindungen als Heterobestandteile in Peptiden VI," Chem. Ber. 98, 3451 (1965).

792—Nikolov, T. K., and D. Kolev: "Use of the sorption agents Whatman SG41 and Whatman CCM1 for thin layer chromatography," Farmatzija [Sofia] 15, 223 (1965).

793—Nitecki, D. E., and J. W. Goodman: "Synthesis of eight dipeptides and four tripeptides of glutamic acid," Biochemistry 5, 665 (1966).

794—Nitecki, D. E., I. M. Stoltenberg and J. W. Goodman: "Qualitative and quantitative determination of mixtures of amino acids using 2,4,6-trinitrobenzenesulfonic acid," Anal. Biochem. 19, 344 (1967).

795—Nitschké, E.: "DC-Trennung und halbquantitative Bestimmung als Blut-Phenylalanins als Verfahren zur systematischen Prüfung auf PKU bei Kleinstkindern," Z. Klin. Chem. Biochem. 6, 208 (1968).

796—Nordwig, A., and W. F. Jahn: "A collagenolytic enzyme from aspergillus oryzae," Europ. J. Biochem. 3, 519 (1968).

797—Numan, J. A., A. Nauman and S. C. Werner: "Total and free triiodothyronine in human serum," J. Clin. Invest. **46**, 1346 (1967).

798—Oelofsen, W., and C. H. Li: "Adrenocorticotropins XXXVI," J. Am. Chem. Soc. **88**, 4254 (1966).

799—Offer, G. W.: "The N-terminus of Myosin I," Biochim. Biophys. Acta **111**, 191 (1965).

800—Ohno, M., and N. Izumiya: "Synthesis of tyrocidine A," J. Am. Chem. Soc. **88**, 376 (1966).

801—Okafuji, T.: "Studies on free amino acids of the blood in normal infants. 1. A technique for the determination by TLC," Acta Paediat. Japan. **70**, 196 (1966).

802—Oppenheimer, H. L., B. Labousse and G. P. Hess: Implication of an ionizing group in the control of confirmation and activity of chymotrypsin," J. Biol. Chem. **241**, 2720 (1966).

803—Oratz, M., A. L. Burks and M. A. Rothschild: "The determination of free ϵ amino-groups in proteins and peptides," Biochim. Biophys. Acta **115**, 88 (1966).

804—Otsuka, H., and J. Shoji: "Structure of triostin C," Tetrahedron **21**, 2931 (1965).

805—Ouellette, R. P., and J. F. Balcius: "A thin layer for the separation of iodine-containing compounds using binary mixtures of adsorbents," J. Chromatog. **24**, 465 (1966).

806—Pais, M., F. X. Jarreau, X. Lusinchi and R. Goutarel: "Alcaloides peptidiques III," Ann. Chem. **1966**, 83.

807—Pataki, G.: "Anwendung der DC zur Sequenzanalyse von Peptiden. 4. Mitt. Untersuchungen über den Abbau von Peptiden mit Phenylisothiocyanat," J. Chromatog. **21**, 133 (1966).

808—Pataki, G., and E. Strasky: "Direct fluorometry and direct spectrophotometry of natural products separated on thin layer chromatograms," 5th International (IUPAC) Symposium on the Chemistry of Natural Products, London, July 1968, p. 27.

809—Pataki, G., J. Borko and A. Kunz: "The non-destructive detection of amino acids on thin layer chromatograms using I-Fluor-2,4-Dinitrobenzene," Z. Klin. Chem. Biochem. **5**, 458 (1968).

810—Pataki, G., and D. Jänchen: "Quantitative in situ fluorometry of thin layer chromatograms," J. Chromatog. **33**, 391 (1968).

811—Pataki, G.: "N-Terminale Sequenzanalyse von Peptiden mittels Edman-Abbaus: Kombinierte Anwendung der direkten und der subtraktiven Methode," Helv. Chim. Acta **50**, 1069 (1967).

812—Pataki, G., J. Borko, H. Ch. Curtius and F. Tancredi: "Some recent advances in thin-layer chromatography. I. New applications in amino acid peptide and nucleic acid chemistry," Chromatographia **1**, 406 (1968).

813—Pataki, G.: "Some recent advances in thin-layer chromatography. II. Application of direct spectrophotometry and direct fluorometry in amino acid, peptide and nucleic acid chemistry," Chromatographia 1, 492 (1968).

814—Pedersen, B. N.: "Mapping of the free amino acids in the cells of bacteria from genus bacillus by TLC," Kgl. Vet. Land-Bohögskole Areskr. 1965, 144; ref. Chem. Abstr. 63, 4683A (1965).

815—Perkins, H. R.: "Homoserine in the cell walls of plant-pathogenic Corynebacteria," Biochem. J. 97, 3C (1965).

816—Petrovic, S. M., and S. E. Petrovic: "Separation and identification of amino acids on starch thin layers," J. Chromatog. 21, 313 (1966).

817—Pettit, G. R., R. L. Smith and H. Klinger: "Synthesis of 3beta-acetoxy-17beta-(L-arginyl-L-prolyl)amino-5$_a$-androstane," J. Med. Chem. 10, 145 (1967).

818—Pfleiderer, G., and A. Krauss: "Die Wirkungsspezifität von Schlangengift-Proteasen (Crotalus atrox)," Biochem. Z. 342, 85 (1965).

819—Pfleiderer, G., and R. Zwilling: "Ueber eine im physiologischen Bereich native Proteine hydrolysierende Endopeptidase aus dem Darm von Tenebria molitor," Biochem. Z. 344, 127 (1966).

820—Pfleiderer, G., A. Stock, F. Ortandel and K. Mella: "Ueber die Coenzymbindung in der Glutamat-Pyruvat-Transaminase aus Schweineherz," Europ. J. Biochem. 5, 18 (1968).

821—Pharmaceutical Society of Society for Analytical Chemistry: "Recommended methods for the evaluation of drugs. Evaluation of thyroid," Analyst 92, 328 (1967).

822—Phocas, I., C. Youanidis, I. Photaki and L. Zervas: "New methods in peptide synthesis. Part IV," J. Chem. Soc. 1967, 1506.

823—Photaki, I.: "A new synthesis of oxytocin using S-acyl cysteines as intermediates," J. Am. Chem. Soc. 88, 2292 (1966).

824—Pirkle, H., and A. Henschen: "N-terminal sequence of bovine fibrinogen," Biochemistry 7, 1362 (1968).

825—Pisano, J. J., E. Prado and J. Freedman: "Beta-Asp-Gly urine and enzymic hydrolyzates of proteins," Arch. Biochem. Biophys. 117, 394 (1966).

826—Plapp, R., and O. Kandler: "Die Aminosäuresequenz des Asparaginsäure enthaltenden Mureins von Lactobacillus coryniformis und Lactobacillus cellobiosus," Z. Naturforsch. 22B, 1062 (1967).

827—Poduska, K. and J. Rudinger: "Amino acids and peptides LXII," Collection Czech. Chem. Comm. 31, 2938 (1966).

828—Poduska, K.: "Amino acids and peptides LXIII," Collection Czech. Chem. Comm. 31, 2955 (1966).

829—Politt, R. J.: "N-2-4-Dinitro-6-sulphophenyl-amino acids: acid hydrolysis," J. Chem. Soc. 1965, 6198.

830—Pollard, L. W., N. V. Bhagavan and J. B. Hall: "The biosynthesis of Gramicidin S. A new peptide intermediate," Biochemistry **7**, 1153 (1968).

831—Porter, T. H., R. M. Gipson and W. Shive: "Syntheses and biological activities of some cycloalenealanines," J. Med. Chem. **11**, 263 (1968).

832—Porter, T. H., and W. Shive: "DL-2-Indaneglycine and DL-β-trimethylsilylalanine," J. Med. Chem. **11**, 402 (1968).

833—Proulx, P.: "Amino acids in lipids of rat brain," Can. J. Biochem. **43**, 523 (1965).

834—Ragnarsson, U.: "The detection of benzyloxycarbonyl-protected amino acid and peptide derivatives on thin-layer chromatograms," J. Chromatog. **30**, 243 (1967).

835—Rampini, S., et al.: "Hereditäre Hyperglycinämie," Helv. Pediat. Acta **22**, 135 (1967).

836—Rannio, R., and J. Puisto: "Formation of amino acids during the first growth phases of Escherichia Coli," Suomen Kemistilehti **39B**, 72 (1966).

837—Rapoport, G., M. F. Glatron and M. M. Lecadet: "Quantitative determination of the N-terminal amino-acyl residues with DNS-Cl," Compt. Rend. (Série D) **265**, 639 (1967).

838—Ratney, R. S., M. F. Godshalk, J. Joice and K. W. James: "The use of 2,4-dinitro-5-fluoroaniline in the determination of protein structure on a micro scale," Anal. Biochem. **19**, 357 (1967).

839—Reinemann, P.: "Bestimmung und Nachweis von Aminosäuren, Peptiden und Proteinen mit Hilfe der DC. Staatsexamensarbeit i.," Inst. f. Org. Chemie, Justus v. Liebig-Institut d. M. Luther-Univ. Halle-Wittenberg **1966.**

840—Reith, W. S., and B. L. Brown: "Separation of iodotyrosines as ^{3}H-DNP-derivatives by TLC followed by electrophoresis on Sephadex," Biochem. J. **100**, 10P (1966).

841—Reusser, F.: "Biosynthesis of antibiotic U-22,324, a cyclic polypeptide," J. Biol. Chem. **242**, 243 (1967).

842—Ried, W., and D. Piechaczek: "Umsetzung von Benzimidsäureäthylester mit beta, γ, delta und epsilon-Aminosäuren," Liebig's Ann. **696**, 97 (1966).

843—Roberts, M., and S. D. Mohamed: "Detection of formimino-glutamic and urocanic acids in urine by TLC," J. Clin. Pathol. **18**, 214 (1965).

844—Roberts, W. S., J. L. Strominger and D. Söll: "Biosynthesis of the peptidoglycan of bacterial cell walls," J. Biol. Chem. **243**, 749 (1968).

845—Robichon-Szulmajster, H. de: "Régulation du fonctionnement de deux chaines de biosynthèse chez saccharomyces cerevisae: thréonine-méthionine et isoleucine-valine," Bull. Soc. Chim. Biol. 49, 1431 (1967).

846—Rohrlich, M., and Th. Niederbauer: "Aminosäurezusammenset-zung und N-endständige Aminosäuren der nach Hess abgetrennten Mehl-proteine," Z. Lebensm. Untersuch. Forsch. 128, 228 (1965).

847—Roncari, G., Z. Kurylo-Borowska and L. C. Craig: "On the chemical nature of the antibiotic Edeine," Biochemistry 5, 2153 (1966).

848—Rosmus, J., Z. Deyl and M. P. Drake: "Studies on the structure of collagen. I. The sequence analysis of peptides released by pronase," Biochim. Biophys. Acta 140, 507 (1967).

849—Rossetti, V.: "Free amino acids in the pollen of lilium candidum," Ann. Chim. 56, 935 (1966); ref. Chem. Abstr. 66, 397D (1967).

850—Roth, M., and A. Reinharz: "Un substrate pour le dosage chimique de la rénine," Helv. Chim. Acta 49, 1903 (1966).

851—Rothe, M., and T. Toth: "Synthese von N-Animoacyllactamen und -cyclopeptiden," Chem. Ber. 99, 3820 (1966).

852—Roubal, W. T., and A. L. Tappel: "Damage to proteins, enzymes and amino acids by peroxiding lipids," Arch. Biochem. Biophys. 113, 5 (1966).

853—Rovery, M.: "Apport des techniques chromatographiques à la connaissance de la structure des enzymes," Bull. Soc. Chim. France 1967, 3935.

854—Russel, D. W.: "Angolide, a naturally-occurring cyclotetradepsid-peptide with a twelve-membered ring," J. Chem. Soc. 1965, 4664.

855—Sakita, Y.: "Method of isolation of isovalthine in urine," Toho Igakkai Zasshi 11, 137 (1964); ref. Chem. Abstr. 16754C (1965).

856—Sakurada, T.: "Separation of human thyroid hormones by TLC," Tohoku J. Exp. Med. 85, 365 (1965).

857—Sakurada, T.: "Separation and quantitative determination of human thyroid hormones by TLC," Tohoku J. Exp. Med. 88, 367 (1966).

858—Sala, E., H. Maier-Hueser and P. Fromageot: "L-lysine-^{14}C and L-histidine-^{14}C. Preparative chromatographic separation," J. Label. Comp. 2, 391 (1967).

859—Samuels, S., and S. S. Ward: "Aminoaciduria screening by thin-layer high voltage electrophoresis and chromatography on micro-plates," J. Lab. Clin. Med. 67, 669 (1966).

860—Santome, J. A., D. E. Wolfenstein, M. Biscoglio and A. C. Paladini: "Sequence of thirteen amino acids in the C-terminal end of bovine growth hormone," Arch. Biochem. Biophys. 116, 19 (1966).

861—Sarges, R., and B. Witkop: "Gramicidin A. V.," J. Am. Chem. Soc. 87, 2011 (1965).

862—Sarges, R., and B. Witkop: "Gramicidin VII," J. Am. Chem. Soc. 87, 2027 (1965).

863—Sarges, R., and B. Witkop: "Gramicidin A VIII," Biochemistry **4**, 2491 (1965).

864—Schally, A. V., and J. F. Barrett: "Isolation and structure of a hexapeptide from pig neurohypophysis," Biochim. Biophys. Acta **154**, 595 (1968).

865—Schildknecht, H., and K. Gessner: "Strukturaufklärung organischer Verbindungen durch Elektronenbrenzen. V. Mitt. Elektronenbrenzen von Alanin, Asparaginsäure, Prolin und Indolyl-3-essigsäure," Z. Anal. Chem. **227**, 338 (1967).

866—Schmer, G.: "Quantitative Bestimmung von $10\text{-}^4\mu$ mol N-terminal amino acids from immunoglobulines by labelling with DNS-Cl," Z. Physiol. Chem. **348**, 199 (1966).

867—Schmer, G., and G. Kreil: "Fingerprints of DNS-labeled protein digests on a millimicromole scale," J. Chromatog. **28**, 458 (1967).

868—Schmitz, H., and H. D. Wilms: "Die Miller'schen Versuche zur chemischen Evolution im Schulunterricht," Math. Naturwiss. Unterricht **19**, 318 (1966/67).

869—Schoellmann, G., and E. Fischer, Jr.: "A collagenase from Pseudomonas aeruginosa," Biochim. Biophys. Acta **122**, 557 (1966).

870—Schorn, H., and C. Winkler: "DC zur Analyse von Schilddrüsenhormonen," J. Chromatog. **18**, 69 (1965).

871—Schultz-Haudt, S. D., J. Aarli, A. L. Nilsen and O. Unhjem: "Hydroxyproline-containing glycopeptides of some human and animal tissues," Biochim. Biophys. Acta **101**, 292 (1965).

872—Schwabe, C., and G. Kalnitsky: "A peptidhydrolase from mammalian fibroblasts," Biochemistry **5**, 158 (1966).

873—Schwartzmann, L., J. Crawhall and S. Segal: "Incorporation of amino acids into a lipid fraction of kidney cortex," Biochim. Biophys. Acta **124**, 62 (1966).

874—Schwyzer, R., and P. Sieber: "Die Totalsynthese des beta-Corticotropins (adrenocorticotropes Hormon: ACTH)," Helv. Chim. Acta **49**, 134 (1966).

875—Scoffone, E., R. Rocchi, F. Marchiori, A. Marzotto, A. Scatturini, A. Tamburo and G. Vidali: "Synthesis of peptides analogous to the N-terminal eicosapeptide of ribonuclease A. Part IV. Synthesis of the enzymatically active orn[10], and orn[10], glu[10]-eicosapeptides," J. Chem. Soc. **1967**, 606.

876—Scoffone, E., R. Rocchi, F. Marchiori, L. Moroder, A. Marzotto and A. Tamburo: "Synthesis of peptide analogs of the N-terminal eicosapeptide sequence of ribonuclease A VII," J. Am. Chem. Soc. **89**, 5450 (1967).

877—Seiler, N., and M. Wiechmann: "Die Bestimmung der γ-Aminobuttersäure im 10-[11]-Mol-Bereich als Dansyl-Derivat," Z. Physiol. Chem. **349**, 588 (1968).

878—Shapiro, O., and A. Gordon: "An improved method for separation of radioactive thyroid hormone metabolites by TLC," Proc. Soc. Exp. Biol. Med. **121**, 577 (1966).

879—Shellard, E. J., and G. H. Jolliffe: "The effect of glycerol on the rate of movement of some amino acids on silica gel thin layers," J. Chromatog. **26**, 503 (1967).

880—Shellard, E. J., and G. H. Jolliffe: "The identification of some amino acids in the presence of 50% glycerol, on silica gel thin layers," J. Chromatog, **31**, 82 (1967).

881—Shields, J. E., and H. Renner: "Synthesis of the carboxy-terminal heptapeptide sequence of bovine pancreatic ribonuclease," J. Am. Chem. Soc. **88**, 2304 (1966).

882—Siebert, G., A. Schmitt and R. von Malortle: "Reinigung und Eigenschaften von Dorschmuskel-Kathepsin," Z. Physiol. Chem. **342**, 20 (1965).

883—Sjöholm, I., and G. Ryden: "Half-life of oxytocin and lysine-vasopressin in blood of rat at different hormonal stages," Arch. Pharm. Suecica **4**, 23 (1967).

884—Skrabka-Blotnicka, T.: "TLC separation of 4-nitropyridylium N-oxide derivatives of amino acids," Chem. Analit. [Warszawa] **12**, 351 (1967).

885—Sluyterman, L. A.: "The effect of methanol, urea and other solutes on the action of papain," Biochim. Biophys. Acta **143**, 187 (1967).

886—Smith, I., L. J. Rider and R. P. Lerner: "Chromatography of amino acids, indoles and imidazoles on thin layers of Avicel and cellulose and on paper," J. Chromatog. **26**, 449 (1967).

887—Southern, E. M., and D. N. Rhodes: "Radiation-induced optical inversion of L-glutamate residues in poly-a-L-glutamic acid," Biochem. J. **97**, 14C (1965).

888—Steiner, L. A., and R. R. Porter: "The interchain disulfide bonds of a human pathological immunoglobin," Biochemistry **6**, 3957 (1967).

889—Stelakatos, G. C., A. Paganou and L. Zervas: "New methods in peptide synthesis III," J. Chem. Soc. **1966**, 1191.

890—Stepanov, V. M., and Y. I. Lapuk: "TLC of 3-methyl-2-thiohydantoins of amino acids," Zhur Obshei Khim. **36**, 40 (1966).

891—Stepanov, V. M., V. I. Ostovslavskaya, V. F. Krivtzov, G. L. Muratova and E. D. Levin: "Characteristic features of the C-terminal fragment of hog pepsin," Biochim. Biophys. Acta **140**, 182 (1967).

892—Stouffer, J. E., P. I. Jaakonmäki and T. J. Wenger: "Gas-liquid chromatographic separation of thyroid hormones," Biochim. Biophys. Acta **127**, 261 (1966).

893—Stransser, H. R., C. R. Schleifer, J. O. Malbica and R. A. Busi: "Studies on human myoglobin by TLC, disc electrophoresis and amino acid analysis," Bull. N.J. Acad. Sci. 11, 15 (1966).

894—Studer, R. O., W. Lergier and K. Vogler: "Die Synthese von Circulin A," Helv. Chim. Acta 49, 974 (1966).

895—Suga, T., I. Ohata, H. Kumaoka and M. Akagi: "Studies on mercapturic acids. Investigation of glutathione-conjugating enzyme by TLC," Chem. Pharm. Bull. [Tokyo] 15, 1059 (1967).

896—Sullivan, G., L. R. Brady and V. E. Tyler: "Identification of a- and beta-amanitin by TLC," J. Pharm. Sci. 54, 921 (1965).

897—Suran, A. A.: "N-terminal sequence of horse spleen apoferritin," Arch. Biochem. Biophys. 113, 1 (1966).

898—Tada, K., Y. Wada and T. Arakawa: "Hypervalinemia. Its metabolic lesion and therapeutic approach," Amer. J. Diseases Children 113, 62 (1967).

899—Tamura, Z., T. Tanimura and H. Yoshida: "Recovery of biologically active peptides from their DNS-derivatives," Chem. Pharm. Bull. [Tokyo] 15, 252 (1967).

900—Tanaka, K., and K. J. Isselbacher: "Isolation and identification of N-isovaleryglycine from urine of patients with isovaleric acidemia," J. Biol. Chem. 242, 2966 (1967).

901—Tancredi, F., and H. Ch. Curtius: "DC von Aminosäuren und Aminen im Stuhl," Z. Klin. Chem. Biochem. 5, 106 (1967).

902—Tannincho, H., and C. B. Anfinsen: "Amino acid sequence of an extracellular nuclease of staphylococcus aureus," J. Biol. Chem. 241, 4366 (1966).

903—Tar, D. F. de, and T. Vajda: "Sequence peptide polymers II," J. Am. Chem. Soc. 89, 998 (1967).

904—Taschner, E., and B. Rzeszotarska: "Aminosäuren und Peptide XIII," Liebig's Ann. 690, 177 (1965).

905—Tettamanti, G., and W. Pigman: "Purification and characterization of bovine and ovine submaxillary mucins," Arch. Biochem. Biophys. 124, 41 (1968).

906—Thoai, N. van, F. Regnouf and A. Olumucki: "Isolement d'un peptide phosphoré et quanidique, l'aspartyllombricine des muscles de Bonellia viridis," Bull. Soc. Chim. Biol. 49, 805 (1967).

907—Thompson, J. E., and M. Spencer: "Ethylene production from β-alanine by an enzyme powder," Can. J. Biochem. 45, 563 (1967).

908—Thornber, J. P., and J. M. Olson: "The chemical composition of a crystalline bacterio-chlorophyll-protein complex isolated," Biochemistry 7, 2242 (1968).

909—Thornton, D. R., and M. H. Fox: "The free amino acid pools of two aquatic hyphomycetes," Experientia 24, 393 (1968).

910—Tinelli, R.: "Séparation des dérivés dinitrophénylés des acides di-amino-pimélique, aspartique et glutaminique par ccm de silice," Bull. Soc. Chim. France **1966**, 348.

911—Tinelli, R.: "Etude de l'alcolyse de parois de listeria monicyto-genes," Compt. Rend. **1965**, 4265.

912—Tinelli, R.: "Etude de la composition du glycopeptide de parois de bactéries Gram + par une microtechnique de ccm," Bull. Soc. Chim. Biol. **48**, 182 (1966).

913—Tipper, D. J., J. L. Strominger and J. C. Ensign: "Structure of the cell wall of staphylococcus aureus, strain Copenhagen VII," Biochemistry **6**, 906 (1967).

914—Tobita, T., and J. E. Folk: "Chymotrypsin C. III. Sequence of amino acids around an essential histidine residue," Biochim. Biophys. Acta **147**, 15 (1967).

915—Toczko, K., and Z. Szweda: "Quantitative determination of PTH-amino acids in thin layer chromatograms," Bull. Acad. Polon. Sci., Sér. Sci. Biol. **14**, 757 (1966).

916—Tometsko, A. M., and N. Delihas: "The application of HVTLE to the separation of peptides, nucleotides and proteins," Anal. Biochem. **18**, 72 (1967).

917—Troughton, W. D.: "Separation of urinary amino acids by TLHVE and TLC," Techn. Bull. Regist. Med. Techn. **36**, 137 (1966).

918—Tschesche, R., R. Welters and H. W. Fehlhaber: "Scutianin, ein cyclisches Peptid-Alkaloid aus Scutia buxifolia Reiss," Chem. Ber. **100**, 323 (1967).

919—Ugolev, A. M., and R. I. Kooshuck: "Hydrolysis of dipeptides in cells of the small intestine," Nature **212**, 859 (1966).

920—Vercellotti, J. R., and A. E. Luetzow: "Beta-elimination of gly-coside monosaccharide from a 3-0-(2-amino-2-deoxy-glucopyranosyl) serine. Evidence for an intermediate glycoprotein hydrolysis," J. Org. Chem. **31**, 825 (1966).

921—Veronese, F. M., A. Fontana, E. Boccu and C. A. Benassi: "Pep-tidi della chinurenina," Gazz. Chim. Ital. **97**, 321 (1967).

922—Vilkas, E.: "Analogie des structures des mycosides C_2 de Myco-bacterium avium et C_6 de Mycobacterium butyricum," Compt. Rend. **1966** (Serie C), 786.

923—Vilkas, E., A. Rojas, B. C. Das, W. A. Wolstenholme and E. Led-erer: "Structure du mycoside C_7 peptido-glycolipide de mycobacterium butyricum," Tetrahedron **22**, 2809 (1966).

924—Vilkas, E., A. Rojas and E. Lederer: "Sur un nouvel acide aminé naturel, la N-méthyl-O-méthyl-L-sérine isolée des mycosides de myco-bacterium butyricum et mycobacterium avium," Compt. Rend. **1965**, 4258.

925—Villanueva, V. R., M. Barbier, C. Gros and E. Lederer: "Sur l'é-thionine produite par Escherichia Coli B," Biochim. Biophys. Acta 130, 329 (1966).

926—Villanueva, V. R., and M. Barbier: "Séparation d'aminoacides soufrés par chromatographies sur papier et sur couches minces," Bull. Soc. Chim. France 1967, 3992.

927—Viscontini, M., and H. Bühler: "Ueber Pyrrolizidinchemie 7. Mitt.," Helv. Chim. Acta 49, 2524 (1966).

928—Vogler, K.: "Synthesis of polymyxin B$_1$," Acta Chim. Acad. Sci. Hung. 44, 143 (1965).

929—Vogler, K., R. O. Studer, P. Lanz, W. Lergier and E. Böhni: "Synthese von Polymyxin B$_1$," Helv. Chim. Acta 48, 1161 (1965).

930—Vogler, K., et al.: "Synthese von Colistin A (Polymyxin E$_1$)," Helv. Chim. Acta 48, 1371 (1965).

931—Vogler, K., et al.: "Synthese von All-D-Val^5-Angiotensin II-Asp^1-beta-Amid," Helv. Chim. Acta 48, 1407 (1965).

932—Vogler, K., P. Lanz, W. Lergier and W. Haefely: "Synthesen von Bradykinin-Analogen mit D-Aminosäuren," Helv. Chim. Acta 49, 390 (1966).

933—Voigt, S., M. Solle and K. Konitzer: Dünnschicht-chromatographische Abtrennung von γ-Aminobuttersäure aus Hirnextrakten," J. Chromatog. 17, 180 (1965).

934—Wada, Y.: "Idiopathic Hypervalinemia," Tohoku J. Exp. Med. 87, 322 (1965).

935—Wakton, P. L.: "The hydrolysis of N-acetylglycyl-L-lysine methyl ester by urokinase," Biochim. Biophys. Acta 132, 104 (1967).

936—Waley, S. G.: "Structural studies of a-crystallin," Biochem. J. 96, 722 (1965).

937—Walker, W. A., and M. Bark: "Separation of urinary and plasma amino acids by two-dimensional tl-electrophoresis and chromatography," Clin. Chim. Acta 13, 241 (1966).

938—Wallenfels, K., and C. Gölker: "Untersuchungen über milchzukkerspaltende Enzyme," Biochem. Z. 346, 1 (1966).

939—Wallis, M.: "A C-terminal sequence from ox growth hormone," Biochim. Biophys. Acta 115, 423 (1966).

940—Walter, R., and I. L. Schwartz: "5-Valine-oxytocin and 1-de-amino-5-valine-oxytocin," J. Biol. Chem. 241, 5500 (1966).

941—Walter, R., and V. du Vigneaud: "6-Hemi-L-selenocystine-oxytocin and 1-deamino-6-hemi-L-selenocystine-oxytocin," J. Am. Chem. Soc. 87, 4192 (1965).

942—Walter, R., and V. du Vigneaud: "1-Deamino-1,6-L-selenocystine-oxytocin," J. Am. Chem. Soc. 88, 1331 (1966).

943—Walter, W., H. P. Harke and R. Polchow: "Das Verhalten von Glykokoll, Alanin, a-Aminobuttersäure, Leucin, Phenylalanin und Asparaginsäure unter hydrothermalen Bedingungen," Z. Naturforsch. **22B**, 931 (1967).

944—Wang, K. T., and J. M. Huang: "Polyamide layer chromatography of DNP amino acids," Nature **208**, 281 (1965).

945—Wang, K. T., and I. S. Wang: "Preparative polyamide layer chromatography," J. Chromatog. **24**, 458 (1966).

946—Wang, K. T., and I. S. Wang: "Chromatographic identification of dinitrophenylamino acids on polyester film supported polyamide layers," J. Chromatog. **27**, 318 (1967).

947—Wang, K. T., I. S. Wang, A. L. Lin and C. S. Wang: "Polyamide layer chromatography of phenylthiohydantoins (PTH) of amino acids," J. Chromatog. **26**, 323 (1967).

948—Wang, K. T., K. Y. Chen and B. Weinstein: "Amino acids and peptides. XV. Separation of N-benzyloxycarbonyl amino acids and esters," J. Chromatog. **32**, 591 (1968).

949—Weicker, H.: "Mikro-Methode zur chromatographischen Bestimmung von Aminosäuren," Z. Klin. Chem. Biochem. **6**, 221 (1968).

950—Waxdal, M. J., W. H. Konigsberg and G. M. Edelman: "The covalent structure of a human γG-immunoglobulin III," Biochemistry **7**, 1967 (1968).

951—Waxdal, M. J., W. H. Konigsberg, W. L. Henley and H. M. Edelman: "The covalent structure of a human γG-immunoglobulin II," Biochemistry **7**, 1959 (1968).

952—Weinert, H., H. Masui, I. Radichevich and S. C. Werner: "Materials indistinguishable from iodotyrosines in normal human serum and human serum albumin," J. Clin. Invest. **46**, 1264 (1967).

953—West, C. D., A. W. Wayne and V. J. Chaure: "TLC for thyroid hormones," Anal. Biochem. **12**, 41 (1965).

954—Weygand, F., W. Steglich, J. Bjarnason, R. Akhtar and N. M. Khan: "Readily cleaved protective groups for acid amide functions I," Tetrahedron Letters (29) **1966**, 3483.

955—Weygand, F., P. Huber and K. Weiss: "Peptidsynthesen mit symm. Anhydriden I," Z. Naturforsch. **22B**, 1084 (1967).

956—Wheatley, V. R., G. Lipkin and T. H. Woo: "Lipogenesis from amino acids in perfused isolated dog skin," J. Lipid Res. **8**, 84 (1967).

957—Whitney, J. G., and E. A. Grula: "A major attachment site for D-serine in the cell wall mucopeptide of Micrococcus Lysodeicticus," Biochim. Biophys. Acta **158**, 124 (1968).

958—Wieland, T., G. Lüben, H. Ottenheim, J. Faesel, J. X. de Vries, W. Konz, A. Prox and J. Schmid: "Antamanid. Seine Entdeckung,

Isolierung, Strukturaufklärung und Synthese," Angew. Chem. **80,** 209 (1968).

959—Wieland, Th.: "Peptides of amanita phalloides," Pure Appl. Chem. **9,** 145 (1964).

960—Wieland, Th., H. Schiefer and U. Gebert: "Giftstoffe von Amanita verna," Naturwissenschaften **53,** 39 (1965).

961—Wieland, Th., and H. Bende: "Chromatographische Trennung einiger diastereomerer Dipeptide und Betrachtungen zur Konformation," Chem. Ber. **98,** 504 (1965).

962—Willemot, J., and G. Parry: "Recherches rapides de traces d'arginine dans l'orithine et ses sels et vice versa par ccm," Ann. Pharm. Franc. **24,** 209 (1966).

963—Wohnlich, J. J.: "Quelques nouveaux aspects de la chromatographie sur verre sablé," Chrom. Symp. III Bruxelles, Presses Académiques Européennes, **1964,** p. 225.

964—Wolman, Y., and Y. S. Kausner: "Detection of tert.-butyloxycarbonyl derivatives on paper and thin-layer chromatograms," J. Chromatog. **24,** 277 (1966).

965—Wood, E. J., and W. H. Bannister: "The effect of photooxidation and histidine reagents on murex trunculus haemocyanin," Biochim. Biophys. Acta **154,** 10 (1968).

966—Woods, K. R., and K. T. Wang: "Separation of dansyl-amino acids by polyamide layer chromatography," Biochim. Biophys. Acta **133,** 369 (1967).

967—Wooley, D. W.: "Solid-phase synthesis of a hexapeptide and test of its chymotryptic activity," J. Am. Chem. Soc. **88,** 2309 (1966).

968—Yamamoto, A., and H. Tsukamoto: "Isolation and identification of N-acety-glucosamine-asparagin," Chem. Pharm. Bull. [Tokyo] **13,** 1046 (1965).

969—Yamashiro, D., D. Gillessen and V. du Vigneaud: "Oxytoceine and deamino-oxytoceine," Biochemistry **5,** 3711 (1966).

970—Yonemitsu, O., P. Cerutti and B. Witkop: "Photoreductions and photocyclisations of tryptophane," J. Am. Chem. Soc. **88,** 3941 (1966).

971—Yoshida, T.: "A simple method for detection of urinary homocystine," Tohoku J. Exp. Med. **91,** 31 (1967).

972—Young, D. S.: "Systematic screening for abnormal urinary constituents," Clin. Chem. (N.Y.) **14,** 418 (1968).

973—Zachmann, M., P. Tocci and W. L. Nyhan: "Occurrence of γ-aminobutyric acid in human tissues other than brain," J. Biol. Chem. **241,** 1355 (1966).

974—Zahn, H., W. Dahno and B. Gutte: "Eine neue Synthese der A-Kette des Schafinsulins und deren Vereinigung mit natürlicher B-Kette zu kristallinem, vollaktivem Insulin," Z. Naturforsch. (B) **21B**, 763 (1966).

975—Zappi, E., and G. Hoppe: "Qualitative Veränderungen verschiedener jodierter phenolischer Aminosäuren im Laufe einer alkoholischen Extraktion aus wässrigen Lösungen," Z. Klin. Chem. Biochem. **6**, 105 (1968).

976—Zappi, E.: "Application of the FFCA-reaction for the detection of thyroid hormones and iodinated derivatives to TLC," J. Chromatog. **31**, 241 (1967).

977—Zappi, E., and G. Hoppe: "Serumextraktion und dünnschicht-chromatographische Trennung zirkulierender Jodamino-säuren," Z. Klin. Chem. Biochem. **5**, 209 (1967).

978—Zappi, E.: "Group separation of an aqueous solution of some iodinated amino acids and derivatives by means of solvent extraction," J. Chromatog. **30**, 611 (1967).

979—Zwilling, R., and G. Pfleiderer: "Eigenschaften der a-Protease aus dem Gift von Crotalus atrox," Z. Physiol. Chem. **348**, 519 (1967).

SCHEME I

Position of amino acids in Solvent No. 42 (first dimension) in the order of increasing R_f-values; below, R_r-values of amino acids in solvents numbers 43, 44 and 45 (second dimension). Compare Fig. 64 and text. The abbreviations are listed in Table 14. isoAbu = α-Aib; β-Abu = β-aminobutyric acid; β-isoAbu = β-Aib; pAhip = p-aminohippuric acid; Can = canavine sulfate; Dabu = α, γ-diaminobutyric acid; Dapim = α, α-diaminopimelic acid; Dopa = dihydroxyphenylalanine; DiJTyr = diiodotyrosine; Dimecys = dimethylcysteine; Gcy = glycocyamine; Hyglu = hydroxyglutamic acid; Hyval = β-hydroxyvaline; Kn = creatine; Knin = creatinine; Kyn = kynurenine; Lan = lanthionine; MetSO₂ = MetO₂; MetSO = MetO; Phegly = α-phenylglycine; Thyron = thyronine; Thyroy = thyroxine. According to von Arx and Neher [5].

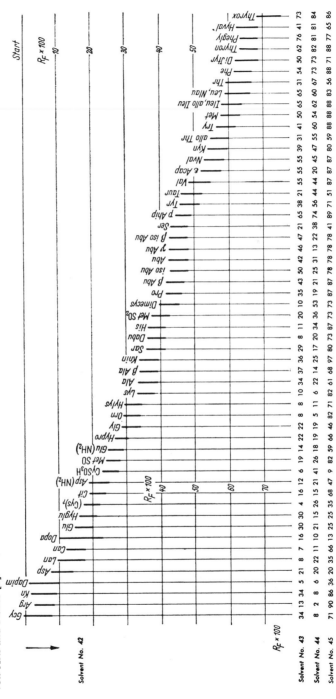

SCHEME II
Scheme of end-group determination according to Sanger

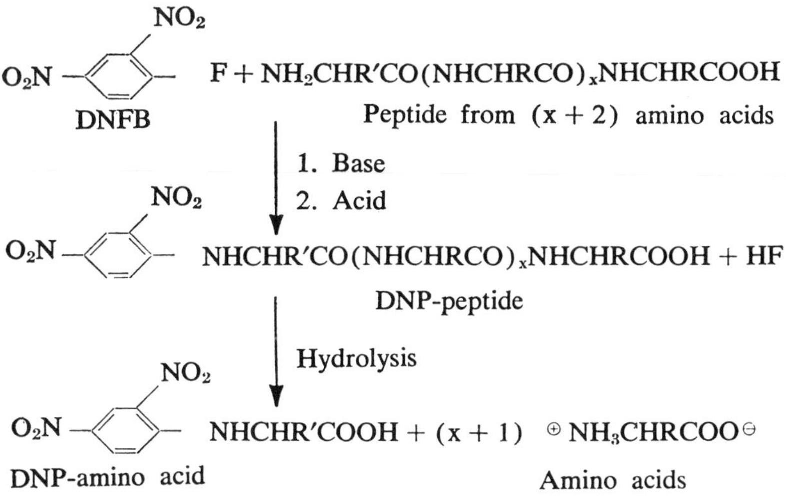

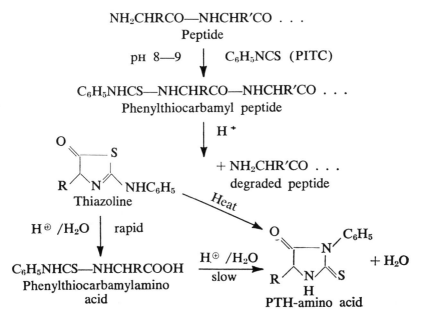

O₂N —⬡— F + NH₂CHR′CO(NHCHRCO)ₓNHCHRCOOH

DNFB   Peptide from (x + 2) amino acids

1. Base
2. Acid

O₂N —⬡— NHCHR′CO(NHCHRCO)ₓNHCHRCOOH + HF

DNP-peptide

Hydrolysis

O₂N —⬡— NHCHR′COOH + (x + 1) ⊕NH₃CHRCOO⊖

DNP-amino acid   Amino acids

SCHEME III
Scheme of the Edman degradation

NH₂CHRCO—NHCHR′CO . . .
Peptide

pH 8—9   C₆H₅NCS (PITC)

C₆H₅NHCS—NHCHRCO—NHCHR′CO . . .
Phenylthiocarbamyl peptide

H⁺

+ NH₂CHR′CO . . .
degraded peptide

Thiazoline

Heat

H⊕/H₂O   rapid

C₆H₅NHCS—NHCHRCOOH
Phenylthiocarbamylamino
acid

H⊕/H₂O
slow

+ H₂O

PTH-amino acid

247

Commercial Suppliers

The following is a list of the United States representatives of some of the foreign commercial suppliers mentioned in this book.

1. C. A. Brinkmann & Co., Inc., Westbury, N.Y., representative of Desaga GmbH, Heidelberg, Germany; of Macherey, Nagel & Co., Duren, Germany; and of E. Merck A.G., Darmstadt, Germany.

2. Camag, Inc., Milwaukee, Wisc., representative of Camag, Muttenz, Switzerland.

3. National Instrument Laboratories, Rockville, Md., representative of Joyce, Loebel & Co.

4. Pharmacia Fine Chemicals, Inc., New York, N.Y., representative of Pharmacia, Uppsala, Sweden.

5. Gallard-Schlesinger Chem. Man. Co., Long Island, N.Y., representative of Serva A.G., Heidelberg, Germany.

6. Alupharm Chemicals, New Orleans, La., representative of Woelm, Eschwege, Germany.

Index

Activity of the Layer, 40
Amino Acids
  buffer for electrophoresis, 88
  chromatography of biological material, 169 ff
  chromatography on aluminum oxide, 74
  chromatography on cellulose, 76 ff
  chromatography on DEAE cellulose, 81
  chromatography on silica gel, 65 ff
  chromatography on silica gel/kieselguhr, 76
  configuration determination, 119
  deproteination of the solution, 170
  desalting, 171
  dinitrophenylation, 128
  electrochromatography, 85 ff
  in biological material, 167 ff
  in blood, 172, 173
  in meat extracts, 179
  in sperm, 189
  in urine, 173
  multidimensional separation, 81
  quantitative determination, 56
  removal of lipoids, 170
  removal of polysaccharides, 170
  R_f-values, 68, 69, 70, 76, 77, 80, 81, 84, 89
  semi-quantitative determination, 56, 176
  separation of fast-moving, 83
  separation of leucine and isoleucine, 73, 82
  solvents, 66, 78
  two-dimensional separation, 71, 73, 75, 79, 82
α-Amino Nitrogen
  in blood, 172
  in urine, 173
Application, 10
  larger quantities of substance, 31
  sensitive substances, 11

Applicators, 3, 4, 5, 6, 7, 8, 27
Automatic Applicator, 6
Autoradiography, 35

Benzyloxycarbonyl Compounds, 100, 102, 103, 105
BN Chamber, 16
  continuous flow chromatography, 19
  horizontal chromatography, 15
  iterating chromatography, 19
  polyzonal chromatography, 23
Buffers for Electrophoresis, 88, 123

Chamber Saturation, 41
Cellulose Layers, 76 ff
Chlorine/Tolidine Reaction, 83, 107
  color data, 91
Chromatography, amino acids, 65 ff
  amino acid butyl esters, 70
  amino acids in biological material, 169, 184
  DNAP amino acids, 144
  DNP amino acids, 131 ff
  DNS amino acids, 145
  instructions, 38
  iodoamino acids, 94
  peptide hydrolysates, 113, 117
  peptides and derivatives, 98
  phenylthiohydantoins, 157 ff
  protein hydrolysates, 113, 117
  PTH amino acids, 157 ff
  TNP alcohols, 164
Column Chromatography
  horizontal, 30
  relation to thin-layer chromatography, 28
Contaminants, influence on the R_f-value, 45
Continuous Flow Chromatography, 19
Creatinine
  in blood, 177
  in urine, 177

DEAE-Cellulose, 77
Densitometer, 60
Detection
  amino acids, 89 ff
  DNAP amino acids, 147
  DNP amino acids, 141
  DNS amino acids, 148
  iodoamino acids, 97
  non-destructive, 22, 36, 92
  peptides, 107
  PTH amino acids, 163
  radioactive substances, 34
Development Distance of the Solvent,
  43
Development of Chromatograms, 10
  ascending, 12
  gradient, 26, 28
  horizontal, 15
  influence on R_f-values, 41
  iterating, 19
  multiplate, 13
  multiple, 16
  polyzonal, 23
  separation-reaction-separation   tech-
    nique, 21
  thick layers, 30
  two-dimensional, 21
  wedge-strip technique, 19
Dimethylaminonaphthalenesulfonyl
    Amino   Acids   (DNS   amino
    acids), 144
  chromatography, 147
  preparation, 144
  separation, 147 ff
  R_f-values, 148
Dinitroaminophenyl   Amino   Acids
    (DNAP amino acids), 144
  chromatography, 147
  preparation, 144
Dinitrophenyl   Amino   Acids   (DNP
    amino acids), 126
  blood, 184
  chromatography, 131 ff
  detection, 141
  ether-soluble, 132 ff
  fermentation charges of grape must,
    191
  photocopy, 142
  photography, 142
  preparation, 127, 128
  R_f-values, 135, 136
  sea water, 191
  separation, 137 ff
  solvents, 132
  sperm, 191
  urine, 184 ff
  water-soluble, 131

Dinitrophenylamino Acid Methyl Es-
    ters, 136
Dinitrophenylpeptide Methyl Ester, 136
Disulfide Bridges, oxidative cleavage,
    115
Double-Labeling, 34

Electrochromatography, 85
Electrophoresis, 33
  amino acids, 85 ff
  amino acids in biological material,
    180-181
  equipment, 33, 34
  peptides, 121 ff
Elution for Quantitative Determination,
    61, 177, 189
End-Group Determination, 130 ff, 162
Evaluation
  qualitative, 39
  quantitative, 56

Fingerprint Technique, 121
Fluorimetry, 59, 60
Folin Reagent, 94
Frontal Analysis of the Solvent, 53

Gas Chromatography, 66, 141
Gradient Technique, 26 ff

Hydrolysis
  acid, 114
  alkaline, 115
  enzymatic, 114
  of peptides, 113
  of proteins, 113
  with ion exchanger, 116

Identification, *see* Detection
Intermediates of Synthesis, 98 ff
Iodoamino Acids
  chromatography, 94 ff
  separation, 96
  solvents, 95
Isatin Reaction, 93
  color data, 91
Isotope Technique, 34

Layer, 65, 76, 77, 159
  activity, 40
  DEAE-cellulose, 77
  manual preparation, 8
  preparation, 3, 30, 77
  pretreatment, 81, 134
  thickness, 4, 6, 31
Layer Application, on glass plates, 3 ff

Martin Relation, 49
  applications of, 55
  chromatography number, 56
  deviations from, 51
  influence of solvent profile, 53
  results, 50
  solvent demixing, 53
Microplate Method, 7
  applicator for, 8
Molecular Weight Determination, 34,
    37
Morin, 108
Multidimensional Chromatography, 21
Multiple Development, 16
  iterating, 19
  R_f-values in, 18

Ninhydrin Reaction, 90
  cellulose layers, 92
  limits of detection, 91
  oxidation with performic acid, 115
  Paulys reagent, 93
  polychromatic, 92
  silica gel layers, 90
  sodium nitroprusside/potassium ferri-
    cyanide reagent, 94

Partition Chromatography, 47
Peptides
  chromatography, 98 ff
  degradation with phenylisothiocya-
    nate, 153
  diastereomeric, 99, 100
  dinitrophenylation, 126
  electrochromatography, 121
  isomers, 99
Peptide Syntheses, control of, 102
Performic Acid Oxidation, 115
Phenylthiohydantoins, *see* PTH Amino
    Acids
Photometry, 59
Preparative Chromatography, 28 ff
  layer preparation, 30
  separation chamber, 32
  substance application, 31
Procedure of Chromatography, 38
Protein Hydrolysates, 113
Proteins
  degradation with phenylisothiocya-
    nate, 156
  dinitrophenylation, 130
PTH Amino Acids
  chromatography, 157 ff
  degradation of proteins and peptides,
    151 ff
  detection, 163
  preparation, 149
  radioactive, 163

PTH Amino Acids (*cont'd*)
  R_f-values, 158, 159
  separation, 161
  solvents, 157

Quantitative Thin-Layer Chromatogra-
    phy, 56
  densitometry, 59
  determination of remission, 59
  double-labeling, 62
  evaluation after elution, 60
  evaluation *in situ*, 57
  fluorimetry, 60
Quantity of Substance Applied, influ-
    ence of the R_f-value, 44

Radiochromatography, 34
  autoradiography, 35
  double-labeling, 34, 62
  evaluation with counter tube, 35
  evaluation with scintillation counter,
    36
  molecular weight determination, 34,
    37
  quantitative amino acid determina-
    tion, 56, 61
Remission Measurement, 59
R_f-Value
  amino acid butyl esters, 70
  amino acids, 68, 69, 70, 76, 77, 80,
    81, 84, 89
  chromatography number, 56
  definition, 39
  determination of, 46, 47
  DNP amino acid methyl esters, 136
  DNP amino acids, 131, 132, 135
  DNS amino acids, 148
  factors influencing, 40
  iodoamino acids, 96
  peptides and derivatives, 99, 100, 102,
    103
  phenylthiohydantoins (PTH amino
    acids), 158, 159
  reproducibility, 47, 48, 40 ff
  TNP amino alcohols, 165
Ribbed Plates, 8, 9

Sakaguchi Reaction, 94
S-Chamber, 14
  additional saturation, 15
Separation Chambers, 12, 13, 14, 15,
    17, 18, 19, 26, 32, 41
Sequential Analysis, 113 ff
Silica Gel, 65, 66
  with indicator, 163
Silica Gel/Kieselguhr, 76

Solvents
demixing, 54
profile, 53
quality, 43
running time, 8
Solvents for Chromatography of
amino acids, 66, 78, 88
DNAP amino acids, 147
DNP amino acids, 132
DNS amino acids, 145, 147
iodoamino acids, 94, 95
peptides and derivatives, 99, 102, 103, 104
PTH amino acids, 157 ff
TNP amino alcohols, 164

Spot Size, 58
determination of, 58
quantitative evaluation, 56
Spotting Template, 12

Technique of Chromatography
ascending, 12
horizontal, 15
Temperature, 45
Trinitrophenylamino Alcohols, 164, 165

UV-Photocopy, 142
UV-Photography, 142

Wedge-Strip Technique, 19, 146

Times Roman is the typeface principally used in this book. The type was set by SSPA Typesetting, Inc., Carmel, Indiana. Printing was by Braun-Brumfield, Inc., Ann Arbor, Michigan, and binding by John H. Dekker & Son, Grand Rapids, Michigan.